Machine Consciousness

Machine Consciousness

edited by

Owen Holland

IMPRINT ACADEMIC

Published in the UK by Imprint Academic
PO Box 200, Exeter EX5 5YX, UK

Published in the USA by Imprint Academic
Philosophy Documentation Center,
PO Box 7147, Charlottesville, VA 22906-7147 , USA

ISBN 0 907845 24 X (paperback)

ISSN 1355 8250 (*Journal of Consciousness Studies*, **10**, number 4–5, 2003)

British Library Cataloguing in Publication Data
A catalogue record for this book is available from the British Library
Library of Congress Card Number: 2002102391

Contents

Contributors

Igor Aleksander (Department of Electrical and Electronic Engineering, Imperial College, Exhibition Road, London SW7 2BT, U.K.) is professor of neural systems engineering at Imperial College, London. He has researched AI systems for thirty years and since 1990 he has worked on the neuromodelling of the brain to understand high level cognitive functions and develop machine consciousness. He was awarded the Outstanding Achievement Medal for Informatics of the Institute of Electrical Engineers in 2000. He has published twelve books, including *Impossible Minds* (1996) and *How to Build a Mind* (2000) and over 200 papers.

Susan Blackmore is a freelance writer, lecturer and broadcaster, and a visiting lecturer at the University of the West of England, Bristol. She has a degree in psychology and physiology from Oxford University (1973) and a PhD in parapsychology from the University of Surrey (1980). Her research interests include memes and the theory of memetics, evolutionary theory, consciousness, and meditation. Her recent books include *The Meme Machine* (1999) which has been translated into nine other languages, and *Consciousness: An Introduction* (2003), the first textbook on consciousness.

Ron Chrisley (School of Computer Science, University of Birmingham, Birmingham, B15 2TT, U.K.) took his DPhil in philosophy at Oxford in 1996 and became a senior lecturer in philosophy at the School of Cognitive and Computing Sciences at the University of Sussex. He is currently Leverhulme Research Fellow in Artificial Intelligence at Birmingham, where he is working on the Cogaff Project.

Rodney Cotterill (Biophysics Group, Building 307, Danish Technical University, DK-2800 Lyngby, Denmark) is professor of biophysics at the Danish Technical University, and was educated in London, Yale and Cambridge. His books include *The Cambridge Guide to the Material World* (1985), *No Ghost in the Machine* (1989), the monograph *Autism, Intelligence and Consciousness* (1994) and most recently *Enchanted Looms: Conscious Networks in Brains and Computers* (1998).

Barry Dunmall (Department of Electrical and Electronic Engineering, Imperial College, Exhibition Road, London SW7 2BT, U.K.) is research fellow in intelligent and interactive systems at Imperial College, London, and has researched systems for neuromodelling and particularly visual awareness in robots for the last ten years. His background is in technical design and project management in industry. He has co-authored several papers on visual awareness and software development for neuromodelling.

Stan Franklin is Dunavant Professor of Computer Science at the University of Memphis, and is co-director of the Institute for Intelligent Systems there. His research is motivated by wanting to know how, human, animal and, particularly, artificial minds work.

For some years he has worked on 'conscious' software agents, autonomous agents modelling a psychological theory of consciousness. His graduate degrees are from UCLA. He has authored or co-authored some eighty papers and one book entitled *Artificial Minds*.

Stevan Harnad (Centre de Neuroscience de la Cognition, Université du Québec à Montréal, CP 8888 Succursale Centre-Ville, Montréal, Québec, Canada H3C 3P8) did graduate work at Princeton and is currently Canada Research Chair at l'Université du Québec à Montréal, researching categorisation, communication and cognition. He founded and edited (until 2002) *Behavioral and Brain Sciences* and also pioneered two online ventures, *Psycoloquy* (an electronic journal sponsored by the APA) and the *CogPrints Electronic Preprint Archive* in the Cognitive Sciences.

Owen Holland (Department of Computer Science, University of Essex, Wivenhoe Park CO4 3SQ, U.K.), senior lecturer in computer science at the University of Essex, has researched in the area of biologically inspired robotics for fifteen years. He has previously held faculty positions in psychology and in electronic engineering, and has worked in universities and research laboratories in the UK, Belgium, Germany, and the USA.

Rodney F. Goodman is a consultant to several advanced technology start-up companies, including Cyrano Sciences, an electronic nose company, of which he is a founder. He is a Fellow of the Institution of Electrical Engineers, and was Professor of Electrical Engineering at the California Institute of Technology until 2001. His current interests include intelligent information processing systems and robotics.

Jesse Prinz (Department of Philosophy, Caldwell Hall, University of North Carolina, Chapel Hill, NC 27599, U.S.A.) joined the faculty at Chapel Hill in 2003. His research interests are in the philosophy of cognitive science, philosophy of language and moral psychology. He has two new books: *Furnishing the Mind* (MIT Press, 2002) and *Emotional Perception* (Oxford University Press, forthcoming).

Aaron Sloman (School of Computer Science, The University of Birmingham, Birmingham B15 2TT, U.K.) took his DPhil in philosophy (Oxford, 1962), which he taught at Hull University and then at Sussex. He spent 1972–3 in Edinburgh as Senior Visiting Fellow, and was converted to 'AI as the best way to do philosophy'. He went on to help found what became the Sussex School of Cognitive and Computing Sciences. He moved to Birmingham in 1991, first as Head of School of Computer Science, and since 1994 as a research professor, working on architectures for human-like agents.

Luc Steels (Vrije Universiteit Brussel. AI Lab. 10G-725. Pleinlaan 2, B-1050 Brussels Belgium) is professor of artificial intelligence at the University of Brussels (VUB) and director of the Sony Computer Science Laboratory in Paris. His work has spanned many areas of artificial intelligence from autonomous robots to natural language. The main focus of his current work is to build models for the origins and evolution of grounded language and meaning in groups of autonomous robots.

William Irwin Thompson is a poet and cultural historian and the author of sixteen books, most recently, *Coming into Being: Artifacts and Texts in the Evolution of Consciousness*. He took his doctorate at Cornell University in 1966, and has taught at Cornell, MIT, and York University in Toronto. From 1972 to 1997 he served as Founder and Director of the Lindisfarne Association (*Lndsfrn@aol.com*). Recently he has designed an evolution of consciousness curriculum for the Ross School in East Hampton, New York, and serves as a consultant to the school's administration and faculty.

Owen Holland

Editorial Introduction

In May 2001, the Swartz Foundation sponsored a workshop called 'Can a machine be conscious?' at the Banbury Center in Long Island (http://www.swartzneuro.org/banbury_2001.cfm). Around twenty psychologists, computer scientists, philosophers, physicists, neuroscientists, engineers, and industrialists spent three days in a mixture of short presentations and long and lively discussions. At the end, Christof Koch, the chair, asked for a show of hands to indicate who would now answer 'Yes' to the question forming the workshop theme. To everyone's astonishment, all hands but one were raised. We had not asked the question at the beginning, and so we did not know if any minds had changed during the workshop, but I think we all realized the significance of this near-unanimous vote: the idea of machine consciousness had progressed from being an interesting philosophical diversion to a real possibility.

Later that year, the editors of the *Journal of Consciousness Studies* agreed that the topic would be suitable for a special issue of the journal, and submissions were invited from some of the Banbury workshop participants, and from others interested in the subject. I am grateful to all of the contributors for their co-operation and collaboration in bringing this collection together, and to the referees for the care with which they undertook their task. Special thanks go to Joseph Goguen, editor-in-chief of the *JCS*, and to managing editor Anthony Freeman for his patience and assistance throughout the project.

Igor Aleksander has spent several years engineering artificial neural systems to investigate and demonstrate various aspects of visual consciousness, particularly those involving imagination and imagery. One consequence is that he has probably spent more time than anyone else discussing and defending the notion that a machine might possess at least some of the attributes of consciousness. In their contribution to this collection, he and **Barry Dunmall** do not present a new neural model, but instead propose an axiomatic framework within which the structural and functional components of conscious systems, natural or artificial, can be identified and tested. They note: 'We deem this to be useful if there is ever to be clarity in answering questions about whether this or the other organism is or is not conscious.' They emphasize that their approach 'is meant to be open-ended', so that others can contribute 'further axiomatic clarifications'. Their current systems, embedded in robots, satisfy only three of their five axioms, and are therefore non-conscious, but within their formalism they are now able to ask: '[G]iven the development or evolution of the remaining two axiomatic mechanisms, what arguments could be used to deny them consciousness?'

Almost all the engineers and computer scientists involved in machine consciousness take a more or less conventional computational or neurally inspired approach, concentrating on the functions associated with cognitive processing. **Susan Blackmore**'s paper should give them pause: she suggests that our distinctively human consciousness centred on an experiencing self is an illusion created by the memes which have shaped our minds, and that the primary requirement for a machine 'to think it was conscious' is the ability to host memes — that is, to possess a capacity for imitation. This is usually low down on the list of cognitive abilities considered for implementation in artifacts (though there are signs that this is changing — see Nehaniv and Dautenhahn, 2002). Put bluntly, Blackmore appears to be saying not just that we might have missed something, but that we might have missed almost everything that matters. In support of her case she advances a wide variety of arguments, ranging from robotic experiments to evidence from meditation; she also identifies some key unanswered questions, asking in particular 'whether artificial meme machines can ever transcend the illusion of self consciousness'.

In his paper on his new project, CyberChild, **Rodney Cotterill** brings together a number of approaches to the problem of machine consciousness. His chosen method is the computer simulation of the brain, body, and environment of a very young infant; the architecture of the child's brain is a close neural model of what he has identified as the relevant parts of the mammalian nervous system; and the strategy is developmental and interactive, in that the child must signal its needs to the experimenter — for example, by crying appropriately — and the experimenter must respond. Cotterill is very open minded: although he has a well developed theory of consciousness, he makes it clear that his current project is broadly investigative, 'searching for the neural correlates of consciousness through computer simulation' rather than explicitly testing any single narrow hypothesis. CyberChild possesses not only a simulated brain and body, but also a simulated metabolism; learning to deal with the contingencies presented by its environment is a matter of life and death, and by implication the approach emphasizes the functional links of emergent consciousness to the well-being of the organism. Although the simulation is necessarily much less complex than the reality it is intended to mimic — as Cotterill puts it, 'In CyberChild, one sees the nervous system pared down to its essentials' — the resultant simplicity offers the advantage that 'If evidence of conscious behaviour does emerge . . . one could be reasonably optimistic that its neural correlates will be detectable.' The work seen in close focus is rooted in biology, but Cotterill expresses the hope that success will constitute 'a step toward realizing the long-cherished dream of creating *Homo siliciens*: consciousness in a computer'.

In many ways, **Stan Franklin**'s work on 'conscious software' offers a real challenge to functionalists. If consciousness is what consciousness does, then his systems may well exceed the requirements, in that they not only mimic successfully the outcomes of conscious processes in some humans (naval despatchers) but they do it in the way that the conscious human brain appears to do it, since their functional components are explicitly modelled on the elements of Baars' global workspace theory (Baars, 1988). Franklin is happy to describe his system IDA as having functional consciousness (in a sense which he defines carefully); in this paper, he goes on to speculate about what more might be said about the system, and what more might be added to extend and perhaps deepen any consciousness that exists in the current version. He is in a privileged position: if IDA is judged not to be conscious because 'she' lacks some functional or structural component X (such as a self), then he and his team may be able in due course to add X. What then? As with Aleksander and Dunmall, who offer their axioms up for extension and modification if anyone feels they are deficient, the onus is on the critic to characterize what is lacking — and if the critic obliges in precise enough terms, it may be possible to correct the deficiency. From a technical point of view, IDA occupies a curious position: she is disembodied, but performs a real-world task in real time. She is also extremely complex. Embodiment and complexity are two key themes raised in other papers in this collection.

In his characteristically vigorous piece, **Stevan Harnad** tackles the issue of how we might know that a machine — an artifact — was conscious, and argues for 'behaviour-based Turing testing' as being necessarily at the root of all our attributions of consciousness, even to other humans. When Johnson remarked of an evening spent in a tavern with company that, 'We had good talk,' Boswell commented, 'Yes sir, you tossed and gored several persons.' Harnad does the same with several ideas, revealing the behavioural nature of the evidence underlying judgments of both the presence and correlates of consciousness. However, advocates of machine consciousness who find their optimism rising as they read the essay may be sobered by the last paragraph, where Harnad reminds us that 'our forward- and reverse-engineering (of Turing indistinguishable robots) can only explain how it is that we can do, not how it is that we can feel.'

Like Cotterill, **Owen Holland** and **Rod Goodman** do not start with consciousness, but hope to end up with it at some future time. And like Blackmore they emphasize a single mechanism — internal modelling — as the possible underpinning of consciousness. (Internal modelling is not imitation, but the two notions are close enough to give food for thought.) Their approach is rooted in robotics; their claim is that a robot able to deal intelligently with the complexities of the real world will have to engage in planning, and that this requirement will inevitably demand the creation of an internal model not just of the world, but of many aspects of the embodied agent itself. They speculate that such an internal agent-model may give rise to some consciousness-like phenomena. Their strategy, like Cotterill's, is developmental, but rather than allowing an entity to modify and extend its own capabilities, they propose to re-engineer the robot themselves, adding and changing whatever is necessary to deal with the progressively more difficult environmental contingencies to which they intend to expose it. Like Aleksander and Dunmall, their starting point is a robot that they claim is definitely not conscious; from there, as they remark, 'The only way is up.'

At the end of his paper, after presenting and defending a version of mysterianism, **Jesse Prinz** offers some advice to engineers: '[E]ngineers should continue trying to model what they can, and they should stop trying to model what they can't. . . . [They] shouldn't fool themselves into thinking they can definitely create conscious machines.' If correct, this is surely good advice; no engineer wants to waste prime-time decades on an impossible quest (that's the territory of the physicist.) The roots of his mysterianism echo the concerns of the single dissenter at the Banbury meeting: the impossibility of being sure that the biological substrate does not contain, at some low level, properties essential for consciousness. After developing his positions on two of the usual objections to machine consciousness, Prinz then goes on to present a theory of consciousness which he uses 'to show that progress in the science of consciousness may offer little help to those who want to engineer consciousness'. Like Harnad, he is concerned that behavioural evidence is at the root of attributions of consciousness, but unlike Harnad, he restricts his concern to dealing with 'inorganic brains', and this is why he is sceptical. Engineers might find a hint of optimism in his statement of what he is sceptical about: 'The problem isn't that it would be impossible to create a conscious computer. The problem is that we cannot know whether it is possible.'

Rather than working from an understanding of consciousness to the construction of a machine to support it, **Aaron Sloman** and **Ron Chrisley** offer a different approach. Arguing that our existing concepts of mind are 'pre-theoretical', they consider progressively more complex information processing architectures based on virtual machines, and use the possibilities offered by various distinct types of architectures both to illuminate some of the different meanings of consciousness, and to suggest more useful and accurate concepts. In the process, they deal with many of the issues which face the designer of any intelligent agent, and provide what is sure to be a useful taxonomy of mechanisms and possible systems. An example of the potential power of their approach can be seen in their discussion of qualia (a topic usually avoided by engineers) within what is essentially an engineering framework. Their paper does not propose the construction of any particular machine — the notions used are abstractions, rather than concrete suggestions for implementation. Abstraction usually implies a degree of simplification, but one of the striking features about their work is the complexity of the architectures they are able to generate.

Luc Steels offers a view of a key feature of consciousness — the inner voice — from the perspective of research into the acquisition and use of language by artificial agents. He expresses his methodological stance as follows: '[W]hatever consciousness "really" is, some of the behaviours often associated with having consciousness can be unravelled, and their information processing foundations understood.' He describes how a community of agents can readily arrive at a shared lexicon by engaging in a robotic 'language game', but how 'the emergence of grammar has turned out to be much more difficult'. This problem was solved by the introduction of a particular information processing strategy — a re-entrant mapping, when output from the language production system is fed back into the language interpretation system. He describes how this enabled the acquisition of a form of grammar by the members of a suitably programmed agent community. However, this re-entrant system (which would not be out of place in one of Sloman and Chrisley's architectures) also has the potential for enabling a range of entirely new processes, including an inner voice that could provide the foundation for the construction and testing of a self-model. Although such a self-model 'is not to be identified with consciousness', he argues that it is part of the conscious experience, and so the development of language in the way he describes 'may have played a crucial role in the origins of consciousness'. Although he is optimistic about the future progress of robotics in capturing some of the information processing aspects of consciousness, he leaves the question of producing first person experience open.

As one might expect, our last contributor, **William Irwin Thompson**, deals with machine consciousness from a rather different point of view, by examining the cultural causes and implications of its pursuit in the West at this particular time. He paints a dark picture. He wrote in 1992, 'There seems little chance of getting out of this century with the same human nature with which we entered it' (Thompson 1992); his concern now is that 'in order to grant consciousness to machines, the engineers first labour to subtract it from humans', and 'the humanistic philosopher of mind . . . finds himself replaced by the robotics scientist'. Interestingly, some of Thompson's observations concerning the mistaken path of 'the mechanists' resonate with some of the papers in this collection. He emphasizes that 'slowness is fundamental to the nature of consciousness', and that the 'slow, sloppy' nature of nervous tissue is responsible for this; both Harnad and Prinz also consider the question of whether the biological substrate may hold some property essential to consciousness. In a passage speculating on the evolution of consciousness, he examines the problems of an organism possessing 'multiple channels of sensory registration'; it is difficult to resist the temptation to locate the creature within Sloman and Chrisley's taxonomy of information processing architectures. However, Thompson is deeply suspicious not only of the methods but also of the motivations of 'the mechanists': he thinks they (and we?), caught up in hypercapitalism, are simply 'hawking their wares'. But he also holds out the hope that monolithic corporate capitalism and American techno-idolatry may not in the end prevail — that they may only delay 'our emergence to an enlightened planetary culture' for a century or two, or even be outcompeted by a new form of capitalism 'wed to information technologies and complex dynamical systems'.

It is now fourteen years since the publication of Leonard Angel's book *How to Build a Conscious Machine* (Angel, 1989) — perhaps the first serious consideration of making a practical assault on the problem of machine consciousness. The book has worn well — some might take this as an indication that progress has been slow. However, it is interesting to note some of the differences between it and Pentti Haikonen's recent book *The Cognitive Approach to Conscious Machines* (Haikonen, 2003). Angel, a university-based philosopher interested in artificial intelligence, wrote from a philosophically-influenced perspective; his main concern was to shed light on 'the traditional mind/body problem'. Haikonen is an engineer working for a major technology company; his text contains system block diagrams, signal flow diagrams, and visual subsystem diagrams, and his preface mentions that he is already working on 'neuron group microchip development for the eventual implementation' of his machines. This gradual shift from the armchair to the laboratory and the workshop can also be seen, I believe, in the present collection. We cannot yet know how fast and how far the enterprise will progress, and how much light it will be able to shed on the nature of consciousness itself, but it seems beyond doubt that machine consciousness can now take its place as a valid subject area within the broad sweep of consciousness studies.

References

Angel, L. (1989), *How to Build a Conscious Machine* (Boulder, CO: Westview Press).
Baars, B.J. (1988), *A Cognitive Theory of Consciousness* (Cambridge: Cambridge University Press).
Haikonen, P.O. (2003), *The Cognitive Approach to Conscious Machines* (Exeter: Imprint Academic).
Nehaniv, C. and Dautenhahn, K. (2002), *Imitation in Animals and Artifacts* (Cambridge, MA: MIT Press).
Thompson, W.I. (1992), *The American Replacement of Nature: The Everyday Acts and Outrageous Evolution of Economic Life* (New York: Doubleday).

Igor Aleksander and Barry Dunmall

Axioms and Tests for the Presence of Minimal Consciousness in Agents

I: Preamble

This paper relates to a formal statement of the mechanisms that are thought minimally necessary to underpin consciousness. This is expressed in the form of axioms. We deem this to be useful if there is ever to be clarity in answering questions about whether this or the other organism is or is not conscious. As usual, axioms are ways of making formal statements of intuitive beliefs and looking, again formally, at the consequences of such beliefs. The use of this style of exposition does not entail a claim to provide a mathematically rigorous formal deductive system. Conventional mathematical notation is used to achieve clarity, although this is elaborated with natural language in an attempt to reduce terseness. In our view, making the approach axiomatic is synonymous with building clear usable tests for consciousness and is therefore a central feature of the paper.

The extended scope of this approach is to lay down some essential properties that should be considered when designing machines that could be said to be conscious. In the broader discussion about the nature of consciousness and its neurological mechanisms, it may seem to some that axiomatisation is premature and continues to beg many questions. However, the approach is meant to be open-ended so that others can build further axiomatic clarifications that address the very large number of questions which, in the search for a formal basis for consciousness, still remain to be answered. Of course, in discussions about consciousness many will also argue that the subject is not one that may ever be formally addressed by means of axioms. The view taken in this paper is 'let's try to do it and see how far it gets'.

The outcome has been positive as it formalises an intuition about which objects have mechanisms which could, potentially, make them conscious. So if some object does not possess some of the major mechanisms indicated by the minimal set it is unlikely that it could be considered to be conscious. For example, a computer *per se* has no mechanisms that are indicated by the axioms and is therefore not conscious. But a virtual machine hosted by an unconscious computer that does obey the axioms might be potentially conscious of, say, a virtual world for which the axioms apply. (This supports a notion voiced by Dennett (1993, p. 210) where he argues that consciousness can be understood as a virtual machine running on a parallel neural computer.) This is visited again at the end of the paper.

II: Scope of the Paper

A set of axioms is proposed which aim to define a minimal set of *necessary* material conditions for cognitive mechanisms that would support consciousness in an agent. For the purpose of this paper 'agent' is used in the sense of an active object that can sense its environment, have a purpose, plan according to that purpose and then choose to act purposefully. An agent can be biological, non- biological and, indeed, virtual (i.e. software running on a host computer). However, casting this material in the domain of physically realisable robot-like machinery is exploited as a hook on which to hang the more general axiomatic theory. The axioms are about perception, imagination, contemplation, planning, acting and emotions. These and other related terms are to be understood in their usual sense in cognitive neuroscience. Hypotheses and predictions that arise from these axioms are examined in the light of findings in biological systems. It will be seen that these indicate that, in order to support consciousness, mechanisms need to be cellular and influenced not only by external perceptual events, but also by the actions needed to gather and organise the perceptual sensations. The collation of axioms and corollaries leads to a final hypothesis about what objects may be said to be conscious. Also, as this theory introduces a variant form of functionalism, the paper concludes with a consideration of some philosophical issues that might be perturbed by this approach.

III: Discerning Consciousness in an Agent

Given are two very similar artificially built agents (A and B) that perform a set of tasks competently. Say that someone claims that one of these (A) is conscious while the other (B) is pre-programmed using conventional computing techniques to cover all eventualities that are likely to be encountered in the performance of the task. Central to our theory is a question we simply call Q.

Q: Is there a set of tests that will substantiate the claim that (A) performs consciously while (B) does not?

To answer this with even a rudimentary working definition D of 'being conscious' makes the point.

D: Being conscious is defined by '*having a private sense*: of an "out-there" world, of a self, of contemplative planning and of the determination of whether, when and how to act'.

(We are fully aware that this will not satisfy those who are concerned with the results of mirror tests on animals which distinguish between conscious organisms which are 'just' conscious and those that are self-conscious. We take a parallel view to Baars, 2001, that all mammals are conscious. For us a reaction to a mirror is merely evidence of an extra ability to *express* what we have called having a private sense of 'self'.)

There appear to be three major approaches for tackling Q: behavioural tests, tests based on an introspective knowledge of consciousness and tests based on a knowledge of mechanism and function. We discard behavioural tests as it has already been said that the two agents perform roughly the same activity. It is also generally accepted that while consciousness may be attributed to agents with an impressive behaviour it cannot be unequivocally inferred from such behaviour. Then, by definition, tests cannot be derived from an introspective feeling of consciousness, as they have to be third-person assessments. While it can be fully accepted that we will never be able to know what it is like to be a conscious agent (Nagel, 1974), the real question is what mechanism does the conscious agent need to possess for *it* to have a sensation of its own presence and affordances for action in its world.

In this paper we take the view that this inevitably implies that it is only through an assessment and analysis of the mechanisms of the organism that we can conclude whether such mechanisms can support consciousness or not within that organism.

IV: The Axiomatic Approach

The axioms of this theory are abstractions drawn from, D, the rudimentary definition of consciousness. In section IX this rudimentary definition is then reviewed in the light of the axioms. As with many axiomatic theories, the real-world problem acts as a stimulator for the origination of axioms which are then examined as abstractions for the generation of hypotheses and theorems that may be reflected back into the real-world domain. Almost incidentally these abstractions may be implemented in the design of artificially conscious agents. In particular, they are candidates for a series of tests directed at the conscious/non-conscious agent question. We leave until the end of this paper setting this approach against extant theories about consciousness. The details of the axiomatic approach are discussed next: we first set out the axioms without explanation and then examine each in greater detail.

Let *A* be an agent in a sensorily-accessible world *S*. For *A* to be conscious of *S* it is *necessary* that:

Axiom 1 (Depiction):
A has *perceptual states* that depict parts of *S*.

Axiom 2 (Imagination):
A has internal *imaginational states* that *recall* parts of *S* or *fabricate S*-like sensations.

Axiom 3 (Attention):
A is *capable of selecting* which parts of *S* to depict or what to imagine.

Axiom 4 (Planning):
A has means of control over imaginational state sequences *to plan actions.*

Axiom 5 (Emotion):
A has additional *affective states* that evaluate planned actions and determine the ensuing action .

In the rest of this paper we show the theoretical and functional characteristics that are implied by these axioms. It should be said that *S* is defined by the senses available to the agent. In humans this adopts the broadest character: it refers not only to the world that can be accessed through all external sensory modalities, but also to physical bodily events that could be internally generated such as heartburn, satiation, thirst, pain, and so on. In an agent equipped only with vision and audition (say) *S* would be only an auditory and visual world.

V: Depiction

Axiom 1: *A has perceptual states that depict parts of S*

We shall pursue the notion that *S* is a conjunction of minimally perceivable events (where an event is understood as a change in the world). This is familiar in vision, for example, where it is possible to build a visual world out of dots as is done in newspapers or television. We further assume that there is an 'energy' involved in any event that can be perceived. In general terms we define a minimally perceivable event δS_j as a minimal element of the sensory world *S* which, were it to have less energy, would not be perceived at all. That is, δS_j indicates the minimum event, the smallest change in the world of which an organism can become conscious. We then write *S* as:

$$S = \boldsymbol{U}_j(\delta S_j)$$

where $\boldsymbol{U}_j$ is an assembly (or, in logic, a conjunction) over all values of *j*, that is of all the minimal events of which the organism could become conscious.

Now we assume that just one minimally perceivable element in *S* has changed. In order for this to be perceived, some functional support in *A* has to change. We define this by means of a corollary related to Axiom 1.

COROLLARY 1.1: To be conscious of δS_j, *it is necessary* for *A* to have a functional support, say $N(\delta S_j)$.

By functional support we mean that the state of some part of the machinery of *A* must have changed. In other words, for an organism to perceive a minimal event in the world requires a related physical change of some state variables of that organism. In neurology $N(\delta S_j)$ is the state of firing of neuronal groups. In the abstract sense $N(\delta S_j)$ is a unique set of values of a number of state variables related to that event. For a different event, that is, a different *j*, a different set of state variables would be affected.

On notation: there could be conditions under which the state variables which support $N(\delta S_j)$ may fail to hold the world-related state. That is, the mechanisms which are necessary for being conscious may not be operating in a conscious mode. In this paper this applies to all cases where the notation $N(x)$ is used.

Assuming a physical world of three spatial dimensions, j has three spatial co-ordinates (x,y,z). It is assumed that $(x,y,z) = (0,0,0)$ is the 'point of view' of the organism (or, in more philosophical terms, the location of the observing self), giving all other world events 'out there' measurements at some distance from this point of view. Clearly j needs to be encoded in $N(\delta S_j)$ in order to retain 'out-thereness' in the inner functional support and this leads to the next corollary.

COROLLARY 1.2: To achieve $N(\delta S_j)$ *it is necessary* for A to have a functional support for the measurement of j. We call this measurement the 'j-referent'.

In living organisms j is often provided by motor action such as the movement of the eyes when foveating on a minimal visual event, or, in some animals, the ability to locate the source of an odour through head movement. In human audition, phase differences and the shape of the ear locate sound. The general point made in corollary 1.2 is that the encoding of the 'where' of a minimal event is as necessary as the qualitative encoding of the sensory event itself in order to give this encoding uniqueness. This leads to an important hypothesis.

HYPOTHESIS 1: *It is sufficient* for each state variable to be indexed by j for the grouping into $N(\delta S_j)$ to be uniquely identified with δS_j, irrespective of the physical location of the variables.

In other words the minimal event is fully encoded internally by the indexing of its state variables irrespective of their physical location. This hypothesis has been discussed elsewhere (Aleksander and Dunmall, 2000). It resolves the 'binding' problem by eliminating it. So in the primate visual system this suggests that neurons in disparate areas such as V3, V4 and V5 may combine to support a single minimal conscious event (e.g. the perception of a small red diamond moving in some direction) with no binding other than the j-indexing of specific neurons. This also leads to a prediction.

PREDICTION 1.1: j-indexed neurons will be found in living conscious organisms.

This prediction should be seen in retrospect as motor-indexed neurons have actually been found in primates since 1989 starting with the work of Galletti and Battaglini (1989), followed by many others. In fact it is quite remarkable how widely spread such discoveries are in the brain: gaze indexing in V3, arm motor indexing in V6, head position indexing in the parietal cortex, and so on.

Given all this it is now possible to define $N(S)$, a *depiction* of S, as the functional support required to be conscious of S, where

$$N(S) = \boldsymbol{U}_j N(\delta S_j)$$

This suggests that all the values of j can be gathered in parallel which somewhat contradicts the practicality of generating values of j through muscular action (such as eye movement). So in living systems, during a perceptual act, internally, j has a sequential nature $j(t)$, *which makes it necessary* for $N(\delta S_{j(t)})$ to have persistence in time (see 'Attention').

In summary, the importance of Axiom 1 is that it suggests a method (the depiction) of internal support for the sensation of an 'out-there' world. It is noted that this strongly suggests a cellularity that determines minimally perceivable events. The support indexes cells to encode the 'out-there' coordinates of minimal events and needs to be compositional through cell interaction, to match the composition of non-minimal events in the world from minimal ones.

VI: Imagination

Axiom 2: *A has internal imaginational states that recall parts of S or fabricate S-like sensations*

Axiom 2 comes from the fact that any organism that is conscious must be conscious of imagined as well as perceived events. Here follow some implied mechanisms that support imagination.

COROLLARY 2.1: The state of a set of cells which supports conscious sensation *has necessarily* at least two components

$$N(C) = N(S)\ ,\ N(I)$$

where $N(I)$ is a j-referenced set of distributed cell states of a dynamic system to which significant states of $N(S)$ are transferred (see corollary 2.2) becoming *attractors* in this set of variables.

Attractors in $N(I)$, in psychological language, would be called recalled memories of past experience, and $N(I)$ in neurological language is a neural memory site. So the process of acquiring experience that may be 'replayed internally' as indicated in the above corollary is a sequence of collective cell states with input $N(S)$ and state $N(I)$. It is suggested that if both are j-indexed they will automatically overlap in consciousness. The next corollary suggests one way in which this transfer may be achieved.

COROLLARY 2.2: One possible way for $N(I)$ to inherit the j-referent is by transfers from $N(S)$ to $N(I)$ through *Iconic learning*

Iconic learning (Aleksander, 1996, p. 122) in a cellular or neural system is the process whereby the states of a network copy the pattern on the input variables of the machine while learning is taking place. Learning is the process whereby each cell associates its own current state with the inputs derived from the outputs of other cells in the system and external inputs.

We note that what we have described is a system of cells each of which determines its state from external input and the states of other cells. In automata theory this is called a *state machine*.

COROLLARY 2.3: Time and sequentiality are the properties of any state machine, therefore $N(I)$ can be the imagination of a sequence.

COROLLARY 2.4: $N(S)$ is not the only input to the state machine. Other inputs could be internal emotions E (see 'Emotion'), internal sensations H (such as pain in living organisms) or verbal input V (in human systems). In very general terms then the state equation for Imagination becomes

$$N(I)' = N(I) \text{ x } N(S) \text{ x } E \text{ x } H \text{ x } V \text{ x } \ldots$$

where $N(I)'$ is the next internal state to $N(I)$ and 'x' indicates a Cartesian product, that is a combined influence of E, H, V ...

This means that imagination sequences may be initiated from a variety of inputs. For example, a sensory state that indicates that a problem has to be solved (a ball is rolling off a table and has to be caught), a state of hunger that needs to be satisfied, or a verbal request has to be understood and acted upon.

In summary, the concept of imagination that is thought to be essential for an agent to be conscious, has been linked to the necessity for a state-machine structure whose states inherit the 'out-thereness' afforded by j-referencing at its controlling inputs.

VII: Attention

Axiom 3: *A is capable of selecting which parts of S to depict or what to imagine.*

Attention is necessary whenever the build-up of $N(S)$ from S cannot be done in parallel. It addresses the question of what generates the j-referent. It is a search process made necessary by any restriction in the degree of parallelism which is available at the input of a system. It is initially driven by agent actions. These actions create the j-referents that enable $N(S)$ to be internally reconstituted. It applies to $N(I)$ as well as a way of bringing detail into consciousness. That is, there is a $N(\delta I_j)$ extracted from $N(I)$ which of necessity parallels the $N(\delta S_j)$ related to $N(S)$.

COROLLARY 3.1: Attention is a search process made necessary by any restriction in the capacity of a sensory input channel. It is initially driven by agent actions to make depiction as complete as possible. These actions create the j-referents that enable $N(S)$ to be internally reconstituted.

The best known example of external attention — i.e. applied to $N(S)$ — is eye movement and foveation. This draws attention to the fact that j-referencing can take place for a group of minimal events simultaneously (i.e. whatever is foveated for one eye position).

COROLLARY 3.2: Attention is a search process that can be context independent or context dependent on S or $N(S)$ or $N(I)$.

COROLLARY 3.3: Attention with respect to $N(I)$, is the inverse process to depictive construction.

Again in primate vision, these three modes can be observed. The neurological system that involves the frontal eye fields and the superior colliculus is the evidence that all three modes indicated by corollaries 3.1, 3.2, 3.3 are possible for perception. This is the system which acts directly on eye muscles, receives input directly from the retina as well as higher visual areas. It moves the eye to areas of high change in the retinal image, or, when it receives inputs from higher areas so it can move the eyes to where important information is predicted to occur.

HYPOTHESIS 2: Inner attention uses the *j*-referent mechanisms that generate $N(S)$ to select parts of $N(I)$.

In other words, selection mechanisms which drive action in external attention (such as the activation of gaze locked cells in vision) are at work when attending to imagined sensations. They (i.e. the *j*-referents) may be a crucial indexing set of signals which position elements of $N(I)$ in the world even when the world is not being perceived. Actuators may or may not be stimulated. If they are not, a clear vetoing mechanism needs to be found.

VIII: Planning

Axiom 4: *A has means of control over imaginational state sequences to plan actions.*

So far $N(I)$ has been described as a passive recall process, possibly indexed by some sort of need. Here we see the same state machine mechanisms working against needs that eventually lead to motion or other forms of action on the part of the agent.

DEFINITION 4.1: Planning is the traversing of trajectories in $N(I)$ under the input control of a need, or a desire, or no input at all, but in all cases leading to the selection of one among several possible outward actions.

COROLLARY 4.1: During exploration, action results in *j*-referents, to have planning, it is necessary for the *j*-referents linked to state trajectories in $N(I)$ also to link to potential action sequences.

The implication of this definition and corollary is that a trajectory in $N(I)$ is a depiction of 'if I do Y will it result in X ?' where X is a need or a desire and Y is a set of *j*-referents on the input of the state machine with the $N(I)$ states. However this allows that X need not be fully defined, leading to something that could be called internal brainstorming. The importance of the role of the *j*-referent is stressed, as 4.1 closes the loop between exploratory mechanisms, the accumulation of experience, and the use of this experience in generating actions. The need for 'gating' (see 'Emotions') is also highlighted, in the sense that plans need to be 'approved' before they lead to action, as will be seen in the next section. This gating has been postulated in primate neurology in cortico-thalamic loops that are 'vetoed' by the action of the anterior cingulate areas (4).

In summary, planning is the primary task for the $N(I)$ state machine, and good planning is a property which aids evolutionary survival.

IX: Emotion

Axiom 5: *A has additional affective states that evaluate planned actions and determine the ensuing action .*

As agreed by many commentators (e.g. Freeman, 1999), emotion is not just a passive change of internal state in response to perceptual events. It is closely linked to the perception–planning–action link.

Plans need to be evaluated, at times very quickly. This implies a mapping from a prediction in $N(I)$. Indeed, to recognise that emotion may be constantly active in consciousness, the following corollary extends corollary 2.1 as follows.

COROLLARY 5.1: The total support of consciousness $N(C)$ contains an emotional element $N(E)$ where this is an evaluation of the states of $N(I)$. That is,

$$N(C) = \mathrm{N}(S), \mathrm{N}(I), \mathrm{N}(E)$$

Where $N(E) = f_e\, N(I)$
f_e being an evaluative function operating on $N(I)$.

We note that f_e may be either inborn (say, fear and pleasure) or more refined through being acquired through experience to a greater or lesser extent (envy, appreciation of beauty).

X: The Consciousness Hypothesis and its Implication

The rudimentary definition D given at the beginning of this paper, uses the words 'having a private sense of …'. The five axioms and their corollaries have suggested the necessity of specific mechanisms without which an agent's consciousness would be incomplete or not existing at all. The axioms all present necessary conditions while it could be argued that they are not sufficient. The key lack of sufficiency lies (as in Chalmers' argument, for example) that an additional factor, say factor F, is required to map from $N(C)$ into $\mathscr{C}$, the ***sensation of being conscious***. In our terminology:

$$\mathscr{C} = F(N(C)),$$

where F is a mapping which (some argue, see XI below) cannot be defined using known science.

In contrast with this, the axiomatic approach suggests the following hypothesis

HYPOTHESIS 3: For every event in $\mathscr{C}$ there is a corresponding event in $N(C)$. That is, in the formulation

$$\mathscr{C} = F(N(C)),$$

the mapping F is one to one making $\mathscr{C}$ a complete description of $N(C)$ and $N(C)$ a complete description of $\mathscr{C}$.

The implication of this is that in order to have a complete theory of consciousness $\mathscr{C}$ it is sufficient to have a complete theory of $N(C)$.

XI: A Philosophical Note

At first sight the 1:1 relationship in hypothesis 3 appears to be classically reductionist. However we have taken care not to eliminate F (some fixed relationship) in this hypothesis by recognising the difference between $\mathscr{C}$ and $N(C)$. We are merely arguing that any theoretical prediction of an event at the subjective level must be formulated at the functional level. Similarly if an explanation is required of a reported event at the subjective level a science of consciousness will find it at the functional level. Therefore a 'science of consciousness' if properly conducted at the functional level will also be a science of the subjective nature of consciousness. In other words, we are not arguing that the sensation of pain is the same as saying that some specific neurons are firing. What we are saying is that the only science that concerns pain is the science that relates to its neural aspect.

We also avoid the expansionist viewpoint of Chalmers (1996) in the sense that we are not saying that $N(C)$ *causes* or *generates* some prime event in some space of fundamental, non-physical subjective $\mathscr{C}$ level. We accept these subjective events but do not accept that they require any science which lies over and above that of the functional level. We therefore accept the special nature of the first person sensation, but argue not that it requires an as yet undiscovered science or link to the functional, but that it is merely a reflection of the functional. In this we present ideas that are similar to those of Velmans' (2000) 'reflexivity' of consciousness. To provide an analogy that plays on the 'reflection' comment, consider the virtual world of reflections in mirrors which only requires the physics of the tangible physical world to explain any event that may be observed in the reflected virtual world.

With respect to physiological theories, we stand far aside from the sequence of Penrose's (1994) and Hameroff's (1994) attacks on computational theories by requiring a quantum source to explain the coexistence of mental states, as when contemplating a problem and visualising its solution simultaneously. We have given a counterexample of this in Aleksander and Dunmall (2000), where an imagined state can coexist with a perceptual state through the concept of the j-referent of corollary 2.1. We would argue that the axiomatic approach in this paper is thoroughly computational.

We do, however, side with the physiological theories of Crick and Koch (1995) in the sense that all sensation has a neural correlate. We differ slightly only in not seeing binding between neural areas as being crucial for superimpositions of neural activities in different brain areas to create a coherent sensation as indicated in Hypothesis 1, which, also, is based on the concept of the j-referent of corollary 2.1.

XII: Conclusion: Consciousness in Machines

In the implication that follows Hypothesis 3, it is necessary to ask what is meant by 'a complete theory of $N(C)$' ? The answer relates back to the five axioms which suggest that it is necessary for certain mechanisms to be present in an agent to create a form (albeit minimal) of consciousness. These mechanisms are recognisable through an analysis of the structural (anatomical in living organisms) and functional (neurological in living organisms) characteristics of a given agent. Such analysis is couched in terms that are computationally realisable. This implies that it is possible to transfer the understood mechanisms into computational agents and thus create systems with a 'machine' consciousness.

Of course, any theories of *N*(*C*), would not be complete at any point in time. Indeed 'completeness' may be something that can only be approached asymptotically. However, whatever the state of understanding of *N*(*C*) is at any point in time, the contention of this paper is that it is machine testable by methods such as computational neuromodelling.

The primary understanding of *N(C)* is likely to come from neurology. An advantage of transferring this to machines, is that hypotheses in neurology can be tested by computation. This was the procedure used in our work on the mechanisms of visual awareness (Aleksander and Dunmall, 2000) and is currently being applied to the discovery of the cause of visual memory deficits among sufferers of Parkinson's disease. Also elements of *N*(*C*) may be directly applied to machinery as is currently being done to enable a mobile robot with vision to develop a sense of visual awareness.

But in addition to what is known in neurology, the transfer of mechanisms into computation allows knowledge to be added to theories of *N*(*C*). A most important feature of a neuromodel is that the content of the conscious experience can be decoded and displayed on the screen (due to Axiom 1 — Depiction). This is a function of the structural knowledge that one has of the connections and coding in the machine. It is unlikely that with the current state of brain scanning systems such a feat could be done for living organisms. It may become, however, become possible in the future at which point the idea that *F* is more than a 1:1 relationship will collapse.

In the preamble to the paper we drew attention to the fact that some commentators on consciousness will see the *C,* which is at the focus of the axioms, as being something different from their concept of consciousness. They argue that this therefore begs the question of whether the mechanisms in this paper are relevant to whatever it is that they call consciousness. However, they would go on to agree that a rock or a computer is not conscious while a human being is. A rock or an unprogrammed computer has none of the axiomatic mechanisms while a bat or a human being might have. We argue therefore that some correlation exists between the presence of consciousness in objects that are reasonably well agreed to be conscious and the existence of the axiomatic mechanisms. However, as mentioned in our preamble, were there to exist a virtual machine in the host computer, and if the presence of the five axiomatic mechanisms could be demonstrated, in what sense could the virtual machine be said to be conscious? The axioms would lead us to believe that the computer could either be conscious of a virtual world resident in the computer or of a real world were the computer be the 'brain' of a robot. An example of a *non-conscious* object may be found in some recently built robots in our laboratory (which sound like not a great achievement). At best one can identify in them the mechanisms that accord to just the first three axioms. Are they conscious? Well, our approach would say that they are not, but given a development or evolution of the remaining two axiomatic mechanisms, what arguments could be used to deny them consciousness? So, here too, we argue that the axiomatic approach issues a challenge to its detractors to provide at least a logical if not axiomatic argument as to why consciousness is something unrelated to the axioms in this paper.

References

Aleksander, Igor (1996), *Impossible Minds: My Neurons, My Consciousness* (London: Imperial College Press).

Aleksander, Igor and Dunmall, Barry (2000), 'An extention to the hypothesis of the asynchrony of visual consciousness', *Proc R Soc Lond B*, **267**, pp. 197–200.

Baars, Bernard (2001), 'Surrounded by consciousness', Paper presented at conference *Towards a Science of Consiousness,* Skovde, Sweden.

Chalmers, David J. (1996), *The Conscious Mind: In Search of a Fundamental Theory* (Oxford Universiy Press).

Dennett, Daniel (1993), *Consciousness Explained* (London: Penguin).

Crick, Francis and Koch, Christopher (1995) 'Are we aware of neural activity in primary visual cortex?', *Nature*, **375**, pp 121–33.

Freeman, Walter J. (1999), *How Brains Make Up Their Minds* (London: Weidenfeld and Nicolson).

Galletti, Carlo and Battaglini Paolo (1989), 'Gaze-dependent visual neurons in Area V3A of monkey prestriate cortex', *Journal of Neuroscience*, **6**, pp. 1112–25.

Hameroff, Stuart (1994), 'Quantum coherence in microtubules: A neural basis for emergent consciousness?' *Journal of Consciousness Studies*, **1** (1), pp. 91–118.

Nagel, Thomas (1974), 'What is it like to be a bat?' *Philosophical Revue*, **83**, pp. 435–50.

Penrose, Roger (1994), *Shadows of the Mind: A Search for the Missing Theory of Consciousness* (Oxford: Oxford University Press).

Velmans, Max (2000), *Understanding Consciousness* (London: Routledge).

Susan Blackmore

Consciousness in Meme Machines

Setting aside the problems of recognising consciousness in a machine, this article considers what would be needed for a machine to have human-like consciousness. Human-like consciousness is an illusion; that is, it exists but is not what it appears to be. The illusion that we are a conscious self having a stream of experiences is constructed when memes compete for replication by human hosts. Some memes survive by being promoted as personal beliefs, desires, opinions and possessions, leading to the formation of a memeplex (or selfplex). Any machine capable of imitation would acquire this type of illusion and think it was conscious. Robots that imitated humans would acquire an illusion of self and consciousness just as we do. Robots that imitated each other would develop their own separate languages, cultures and illusions of self. Distributed seflplexes in large networks of machines are also possible. Unanswered questions include what remains of consciousness without memes, and whether artificial meme machines can ever transcend the illusion of self consciousness.

I am going to set aside some of the major problems facing machine consciousness and concentrate on the question of what sort of machines might acquire human-like consciousness.

The main problem to be ignored is that we do not know how to recognise consciousness in a machine. That is, there is no obvious equivalent of the Turing test for consciousness. I shall define consciousness here in terms of subjectivity; what is sometimes known as 'phenomenal consciousness' (Block, 1995) or 'what it's like to be' (Nagel, 1974). With consciousness being subjective, any objective test, such as any variation on the Turing test, fails to grasp it. You could certainly have a test that shows whether other people *think* a machine is conscious but this is not the same thing, as our eager propensity to attribute feelings and intentions to even the simplest of robots and mechanical toys reveals. Once we start asking whether there is *really* something it is like to be the machine, or whether the world appears a certain way *for* that machine, then our usual tests fail.

In fact we do not know how to recognise consciousness in anything at all. As far as other humans are concerned this is the problem of other minds, but we usually ignore it on the grounds that we think we know what our own consciousness is like and we then extrapolate to others. We cannot do this so easily for other species, hence the problem of animal consciousness, and it is even more difficult with machines.

This first problem is exacerbated by the suspicion that there may be kinds of consciousness utterly different from human consciousness, or indeed from any naturally occurring kind of consciousness. This suggests many interesting lines of thought, but many difficulties too. So I am going to ignore this here, and stick to questions concerning only human-like consciousness. This is quite hard enough to be going on with.

Having put these problems aside I shall ask how we might set about constructing a machine that will have human-like consciousness. I am not a programmer or robot engineer and my purpose is not to discuss the details of construction (I would not be able to do so) but is instead to consider the general principles involved. All theories of consciousness have implications for how consciousness might be artificially created, and pondering these may help us better understand those theories. The theory to be discussed here is that ordinary human consciousness is an illusion created by memes for their own propagation. The implication for machine consciousness is that only a machine capable of imitation would develop a human-like illusion of consciousness.

I shall take the most obvious approach, which is first to ask how the illusion of consciousness comes about in humans, and then use that to ask how it might be artificially created. I shall first make some comments about the general nature of consciousness, then consider how it arose in humans, both during evolution and during individual development, and then see what implications this has for machine consciousness.

Consciousness as Illusion

If we hope (or fear) to make a conscious machine it would be helpful to know what consciousness is. We do not. I shall not claim here to solve the hard problem, or to say what consciousness ultimately is (if anything). Instead I shall argue that ordinary human consciousness is an illusion. Therefore, making a machine that is conscious in the same way as we are means making one that is subject to the same kind of illusion. Before explaining this in more detail I want to distinguish this view from some other major positions on machine consciousness, crudely divided here into three.

1. Machine consciousness is impossible

Among those who have argued that machine consciousness is impossible are dualists, those who believe in a God-given soul as the seat of consciousness, eliminative materialists who do not believe that consciousness exists in the first place, and those who argue that there is something special about biological brains that precludes anything else from having human-like consciousness. This last is particularly confusing but the problems are well known and discussed (Dennett, 1995; McGinn, 1987, pp. 279–88; 1999; Turing, 1950). It is worth noting that Searle, in spite of his theory of biological naturalism (Searle, 1992), does not exclude the possibility of machine consciousness. Rather, he says that any machine could be conscious if it had the same causal properties as a biological brain (Searle, 1997). Since he does not say what those properties are, this is no help in creating artificial consciousness.

Arguments against computational functionalism and good-old-fashioned Strong AI (e.g., Searle's Chinese Room thought experiment), are considered by some to weigh against the possibility of machine consciousness. A more common view is probably that while syntax (or running a formal program) is not sufficient for semantics, and symbols must somehow be grounded in the real world, this can be done by allowing a machine to interact with the world. Developments in neural networks, embodied cognition and situated robotics (e.g., Brooks, 1991; Clark, 1997) suggest the same thing. So this line of argument does not preclude the possibility of conscious machines. Finally, there may be some people who apply to consciousness, rather than to intelligence, what Turing called the 'Heads in the Sand' objection; 'The consequences of machines thinking would be too dreadful. Let us hope and believe that they cannot do so'. (Turing, 1950). None of these arguments provides a good reason for thinking that machine consciousness is impossible.

2. Find consciousness and put it in a machine

Perhaps there is an 'X', or 'extra ingredient', that if we could give it to machines would ensure they were conscious (Chalmers, 1995). McGinn calls the property that would explain consciousness C*, and asks whether it is possible in inorganic materials or not (McGinn, 1999). Some theories of consciousness can be used to derive an 'X' and so to suggest how it could be given to a machine.

I shall not review all the many theories here, but will take just one example as an illustration; the currently popular Global Workspace Theories (Baars, 1988; Dehaene and Naccache, 2001; Dennett, 2001). GWTs equate the contents of consciousness with the contents of a global workspace, likened to a bright spotlight shining on the stage of working memory (Baars, 1988). The GW is a large network of interconnected neurons. Its contents are conscious by virtue of the fact that they are made globally available, or broadcast, to the rest of the system, which is unconscious. I have argued elsewhere that this notion is incoherent and cannot explain consciousness (Blackmore, 2002). GWT depends on the assumption that at any time there is a valid answer to the question 'what is in consciousness now', and that things can meaningfully be said to be either 'in' or 'out' of consciousness, as though consciousness were a container. This is a version of Cartesian materialism (Dennett, 1991) and cannot, I suggest, explain consciousness.

On these theories, 'X' is having a GW. So presumably a machine should be conscious if it is designed with a GW whose contents are broadcast to the rest of its system (see Franklin, this volume). Unfortunately, as mentioned above, even if such machines were built, it would be impossible to test whether they were conscious or not. I can only say that I do not believe that GWT is the way to understand consciousness, and in any case it will have to be tested by other means than making such machines. In the mean time I prefer the third approach, which is to say that consciousness is not what it appears to be.

3. Consciousness is an illusion.

There are several ways of thinking about consciousness as an illusion. Most important is to distinguish them from the view that consciousness does not exist. To say that consciousness is an illusion is to say that it is not what it appears to be. This follows from the ordinary dictionary definitions of 'illusion', for example, 'Something that deceives or misleads intellectually' (Penguin); 'Perception of something objectively existing in such a way as to cause misinterpretation of its actual nature.' (Webster). This point is frequently misunderstood. For example, Velmans (2000) wrongly categorises Dennett's position as eliminativist when it is better described as the view that consciousness is an illusion. I shall explore a version of this position here.

On this view, human-like consciousness means having a particular kind of illusion. If machines are to have human-like consciousness then they must be subject to this same kind of illusion. I shall therefore explore one theory of how this illusion comes about in humans and how it might be created in machines; the theory of memetics.

Human Beings as Meme Machines

Memes are ideas, habits, skills, stories or any kind of behaviour or information that is copied from person to person by imitation (Dawkins, 1976). They range from single words and simple actions to the vast memeplexes (co-adapted meme complexes) of science, art, religion, politics and finance. There are interesting difficulties concerning definitions (Aunger, 2000; Blackmore, 1998), and whether memes can be said to be replicated by means other than imitation, but the essential point is this. When people copy actions or words, those actions or words are copied with variation and then selectively retained and copied again. In other words the actions and words (the memes) fulfil the conditions for being a replicator in a Darwinian evolutionary process (Dawkins, 1976; Dennett, 1995).

This new evolutionary process can only run if the replication process is good enough (has high enough fidelity). Some species of birds, and some cetaceans, can copy sounds with high fidelity, and their songs are therefore memes. But very few other species can imitate at all. Even chimpanzees and orang-utans are, at best, poor imitators and there is much debate over the extent to which they are really able to copy observed behaviours (Dautenhahn and Nehaniv, 2002). Humans appear to be the only species that readily and easily imitates a wide variety of sounds and actions. This suggests that we alone are supporting this second evolutionary process; cultural or memetic evolution. If this is so, human evolution must have taken a very different course from that of other species once we became capable of imitation. I have suggested that human brains and minds were designed by the replicator power of this new process and that this explains why humans are so different from other species (Blackmore, 1999).

There are two aspects of this that are relevant to machine consciousness. First there is how we living humans got to have such large and peculiarly capable brains (the co-evolutionary story). Second is how our individual minds and our sense of self and consciousness are designed by memetic pressures (the developmental story). Both are relevant to the possibility of machine consciousness.

Meme gene co-evolution

The human brain is excessively large by ape standards, and has been extensively redesigned for language (Deacon, 1997). There is no generally accepted theory to explain this but all existing theories have in common the assumption that the ultimate beneficiary is the genes, and that the large brain and capacity for language must have been adaptive from the genes' point of view (Deacon, 1997; Donald, 1991; Dunbar, 1996; Wills, 1993). I have argued, instead, that both were designed by and for the memes in a process called memetic drive.

Once imitation of sufficiently high fidelity occurs, memetic drive works as follows. In a given population of people, memes compete to be copied. People who are especially good at imitation gain a survival advantage by being able to copy the currently most useful memes. Assuming that imitation is a difficult skill requiring extra brain power, this gives an advantage to genes for bigger brains and better imitation. Increasing imitation then provides scope for more competing memes to appear (both useful and harmful ones), and hence there is pressure to be a selective imitator. One effective strategy might be to copy the 'meme founts' — those skilful imitators who pick up currently popular memes and create new ones from the old. Meme founts acquire both status and better opportunities for mating. So they pass on the genes that made them good at propagating *those particular memes*. Memetic drive creates not only bigger brains but brains that are *better adapted to copying the memes that were successful during the previous memetic competition* — whether or not those memes were directly beneficial to people or their genes.

Music and religion are examples. Once people can copy simple sounds, such as humming or drumming, the sounds themselves compete to be copied. People who are best at copying the winning sounds acquire status and a mating advantage. In this way the successful sounds give an advantage to genes for the ability to copy those particular sounds. Similarly with religious behaviours such as rituals and devotions, the winning memes drive brains to become better at imitating *those particular behaviours*. The result is brains that are musical and inclined to religious behaviour.

I have argued that this same process can explain the evolution of language. In general, successful replicators are those with high fidelity, longevity and fecundity. Digitisation of sounds into words may increase fidelity, combining words into novel combinations may improve fecundity, and every improvement leads to increased memetic competition. The people who can best copy the winning sounds have an advantage and pass on the genes that gave them that ability. Gradually, human brains would be driven by the emerging language itself. In most theories of language evolution, the ultimate function of language is to benefit genes. On this theory it is to benefit memes.

Underlying these examples is the general principle that replicators co-evolve with their replication machinery, just as genes must once have co-evolved with their cellular copying machinery. In the case of human evolution, memetic evolution drove the genes to construct better meme-spreading brains. More recent examples include the invention of ever better meme spreading devices from roads and railways to the telegraph, telephone and email. In each case the products copied helped spread the copying machinery which in turn made more products possible and so on. From the memes' point of view the internet is an obvious step in improving meme-copying facilities. As Ridley points out 'memes need a medium to replicate in. Human society works quite well; the Internet works even better' (Ridley, 2003, p. 222). It is in this context that I want to look at the possible development of conscious machines.

Mind design by memes

The second relevant issue is how memes design individual minds; that is, how the design process of evolution unfolds in the case of individual people infected with a lifetime of competing memes.

We spend our lives bombarded by written, spoken and other memes. Most of these are ignored. Some are remembered but not passed on. Others are both remembered and passed on. Some are recombined in novel ways with others to produce new memes. Note that there is much dispute about whether we should use the word 'meme' to apply only to the behaviours themselves, only to the patterns of neural representation (or whatever underlies their storage inside brains), or to both (Aunger, 2000). I shall stick to Dawkins's original definition here, treating memes as 'that which is imitated', or 'that which is copied'. So I shall not distinguish between memes instantiated in books, computers, ephemeral behaviours or human brains, since all can potentially be replicated.

On the memetic hypothesis, human development is a process of being loaded with, or infected by, large numbers of memes. As Dennett (1995) puts it 'Thousands of memes, mostly borne by language, but also by wordless "images" and other data structures, take up residence in an individual brain, shaping its tendencies and thereby turning it into a mind' (Dennett, 1991, p. 254). Language is the mainstay of this process. We have brains specially designed to absorb the language we hear (see above), to deal with grammar, and to imitate the particular sounds of the language(s) we grew up with. By the age of about three years the word 'I' is used frequently and with increasing sophistication. The word 'I' is initially essential to distinguish one physical person from another, but very rapidly becomes used to say things like 'I think', 'I like', 'I want', 'I believe', 'That's mine' and so forth, as though there were a central self who *has* opinions, desires and possessions. In this way a false notion of self is constructed.

There have been very many theories of the formation of this illusory self (Gallagher and Shear, 1999). The difference between other theories and the memetic theory proposed here lies in the question 'Who benefits?'. Previous theories suggest that either the individual person or their genes are the primary beneficiaries; memetic theory suggests that the memes are (Dennett, 1995). I have argued as follows (Blackmore, 1999); once a child is able to talk about his or her self then many other memes can obtain a replication advantage by tagging onto this growing memeplex. For example, saying a sentence such as 'I believe x' is more likely to get 'x' replicated than simply saying 'x'. Memes that can become *my* desires, *my* beliefs, *my* preferences, *my* ideas and so on are more likely to be talked about by this physical body, and therefore stand a better chance of replication. The result is the construction of an increasingly elaborate memetic self. In other words, the self is a vast memeplex; the selfplex (Blackmore, 1999).

The selfplex and the illusion of consciousness

The result of the memetic process described above is that physical, speaking, human bodies use the word 'I' to stand for many different things; a particular physical body; something inhabiting, controlling and owning this body; something that has beliefs, opinions and desires; something that makes decisions; and a subject of experience. This is, I suggest, a whole concatenation of mistakes resulting in the false idea of a persisting conscious self.

The view proposed here has much in common with James's (1890) idea of the appropriating self, and with Dennett's (1991) 'centre of narrative gravity'. There are two main differences from Dennett. First, Dennett refers to the self as a 'benign user illusion', whereas I have argued that it is malign; being the cause of much greed, fear, disappointment and other forms of human suffering (Blackmore, 2000). Second (and more relevant here) Dennett says 'Human consciousness is *itself* a huge complex of memes. . .' (Dennett, 1991, p. 210).

There is reason to question this. Dennett's statement implies that if a person were without memes they would not be conscious. We cannot, of course, strip someone of all their memes without destroying their personhood, but we can temporarily quieten the memes' effects. Meditation and mindfulness can be thought of as meme-weeding techniques, designed to let go of words, logical thoughts and other memetic constructs and leave only immediate sensory experience. The nature of this experience changes dramatically with practice, and it is common for the sense of a self who is having the experiences to disappear. This same selflessness, or union of self and world, is frequently reported in mystical experiences. But far from consciousness ceasing, it is usually described as enhanced or deepened, and with a loss of duality. *If* this experience can justifiably be thought of as consciousness without memes, then there is something left when the memes are gone and Dennett is wrong that consciousness *is* the memes. It might then be better to say that the ordinary human illusion of consciousness is a 'complex of memes' but that there are other kinds of consciousness.

This is, however, a big 'if', and raises all the problems associated with first-person exploration of consciousness (see Pickering, 1997; Varela and Shear, 1999). At present we should not think of this so much as evidence against Dennett's view as a motivation for further research and self-exploration. It might turn out that if meditation is even more deeply pursued and the selfplex is completely dismantled, then all consciousness does cease and Dennett is correct.

The alternative I want to defend here is that memes distort consciousness into an illusion rather than constitute it. On this view the underlying consciousness itself remains unexplained but we can understand the particular nature of ordinary human consciousness in terms of the selfplex. By creating the illusion of self for their own survival and replication, memes are responsible for our false sense that there is always an 'I' having experiences, and for the inherent dualism that bedevils all our attempts to understand consciousness.

On this view many kinds of machine might be conscious, but only a particular kind of machine could be conscious in a human-like, illusory way. It would have to be capable of imitation (otherwise it could not replicate memes) and live in a community of similar meme-sharing machines (otherwise there would be no pressure for memeplexes to form). Such a machine would, if this theory is correct, be a victim of the same illusions of consciousness as we humans are. That is, it would think it had an inner self who was conscious. Ultimately, it would start wondering what consciousness was and trying to solve the hard problem.

We may now consider some implications of this theory for actual machines.

Artificial Meme Machines

There are two kinds of artificial meme machine to consider; those which imitate each other and those which imitate humans.

Machines imitating humans

The strong prediction for machines that imitate humans is that they would come to be conscious in exactly the same way as we do. That is, they would acquire language, social conventions of behaviour, and knowledge of the intellectual world in roughly the same way as we do. In the process of acquiring all these memes, many memeplexes would form, including a selfplex and so they would end up as deluded as we are.

This simple picture is, however, rendered almost completely unrealistic by the fact that humans are not general-purpose copying machines but highly selective imitation devices (Blackmore, 2001). In particular we have complex perceptual, social and communication systems derived from our primate origins (Matsuzawa, 2001), categorisation systems biased towards discriminating certain objects and actions rather than others (Pinker, 1997), dedicated language systems (such as a language acquisition device, innate grammar or language instinct, (e.g., Pinker, 1994), and (if the above theory is correct) specialised capacities to imitate certain behaviours such as music, dance, ritual, art and story telling, rather than others (Blackmore, 1999).

I can only speculate that it is not necessary to construct a machine with exactly these abilities to get the same memetic effect. How similar the abilities must be remains an open question. An analogous question has been faced by those working on the Cog project in trying to design appropriate sensory systems (Brooks *et al.*, n.d.). They have chosen to copy some aspects of human perceptual systems but not all, and presumably this project will discover what differences this makes. However, Cog barely imitates. Cog can point to objects, recognise joint attention and imitate simple behaviours such as head nods, but it was not designed as a meme machine. The position is complicated by the fact that observers often attribute imitation and other abilities to robots, especially 'sociable robots' such as Kismet, even when they do not have them (Breazeal, 2001). If the theory discussed here is correct, then Cog, Kismet and other robots will never acquire a human-like illusion of consciousness unless they are dramatically redesigned to be capable of true imitation learning.

If they were designed that way, or if future robots are capable of imitating human behaviour, then these robots ought to start taking part in human memetic evolution. If they have identical abilities to those of natural humans then the situation will be equivalent to having more people sustaining the evolutionary process. The more interesting (and probably more likely) possibility is that they are sufficiently like us to join in our culture, but sufficiently different to change it. For example (speculating wildly), let us imagine they have faster processing, far larger storage capacity and instant web access; but poorer sensory systems, less agile body movements and less subtle emotions. In this case, the memes they most enjoy acquiring and passing on will differ somewhat from our favourites of gossip, food and sex, but the robots will think they are conscious because they too will go through the process of acquiring human language, using the word 'I', attributing beliefs, desires and possessions to it, and thereby forming a selfplex.

In addition, we humans would probably find ourselves in friendships and other relationships with these creatures. This would affect the entire process of memetic evolution and transform our culture. In the process we would be changed too but, presumably, we would still think we are conscious and so would they. Both kinds of creature would live with the illusion of being a self experiencing stream of consciousness.

At some point these machines will start wondering whether we humans are *really* like them or not. They might propose a 'reverse Turing test' to see whether we are intelligent, but we would certainly fail that. More relevant here, they might try to invent a 'reverse Turing test' to find out whether we are conscious. They would no doubt confront all the familiar problems of subjectivity and other minds. But by the time this happened we would probably already be treating them as conscious beings like ourselves, whether or not we have resolved the problems of consciousness.

Machines imitating each other

One of the most important predictions of the memetic theory of language evolution is that all you need for language to arise is a community of imitating creatures, or meme machines. This is quite different from theories that assume that the capacity for symbolic representation is needed first, such as Donald's (1991) theory of mimesis (note that mimesis is *not* imitation) and Deacon's (1997) crossing of the 'symbolic threshold'. On the memetic theory, the only evolutionary 'turning point' or 'threshold' to be crossed is the ability to imitate. Once imitation is in place, language evolves and reference emerges naturally as the language develops.

I previously suggested that this could be tested by creating a group of imitating robots (copybots) capable of copying each others' sounds (Blackmore, 1999). It turns out that I overlooked the need for shared gaze in creating shared meaning, and wrongly assumed that real copybots could not be built for some time. In fact, they have already been built and the prediction is confirmed.

Steels (2000) describes language as a living system that is continuously evolving and adapting, and has modelled its origin in groups of autonomous distributed agents, using both simulations and robots interacting in the real world. In a series of experiments, two robots can detect and make sounds. They play an imitation game in which one produces a random sound from its repertoire and the other recognises that sound in terms of its own repertoire and then reproduces it. The first then tries to recognise it again. In similar experiments, De Boer (1997) showed that in such a system the culturally successful sounds propagate, and shared vowel systems emerge through self-organisation. Steels has extended this work to demonstrate how shared meanings can also arise using robots that have simple vision systems, categorisation systems and lexicons, and that look at simple coloured shapes presented in front of them. Together the robots come to agree on sounds that refer to something about the shapes they are looking at. Other experiments investigate the emergence of syntactic structures and grammar (Batali, 1998; Kirby, 2001; Steels, 1998). The general conclusion from this, and related work, is that language can be treated as an evolving system in which both syntax and semantics emerge spontaneously.

This research is in its early stages but a fascinating question is whether such robots will spontaneously invent self reference. I expect them to do so, and to re-enact the whole process of selfplex creation. That is, they will initially use the word for 'I' to refer to a physical robot, then begin to talk about it having beliefs, desires, possessions and so on, and this will in turn provide the opportunity for memes to cluster around the growing self concept. Once this happens these robots will, like the human-imitating robots above, acquire the illusion that they are a conscious self experiencing an objective world.

Note that it will be difficult for us to understand their language. If they were designed to have perceptual and categorisation systems just like ours then their language would presumably have a similar structure to human languages. In this case we would be able to learn their language just as we learn human languages, by immersing ourselves in their world (unless they provided us with textbooks and dictionaries to study). But if their perceptual and categorisation systems were different from ours then their language might be extremely difficult for us to learn. We would find it hard to understand what they were talking about because they would parse the world in different ways and talk about completely different things from us. For this reason we would be less likely to attribute consciousness to them, whether or not they had invented self-reference, spawned selfplexes, and hence suffered the illusion of being a conscious self.

This world would contain two (or more) completely different kinds of language. Whether this would be a stable mixture I do not know.

Future Machines

Neither of the types of machine described above is likely to be built in large numbers for the following reason. Imitation is an extremely crude and low-fidelity method of copying information. It is the best that evolution has come up with for copying between individuals but is easy to improve on with digital technology. For example, text and images are now routinely copied with extremely high fidelity all over the world between relatively crude computers linked by the internet. If we want effective robots it will be better to provide them with the capacity to get what information they need this way than for them to go through the slow and poor-quality process of behaviour imitation or actual speech. Also, skills learned by one could be directly transferred to another. Will such machines be conscious in a human-like way? That depends, according to the theory I have outlined here, on whether the machines' memes can gain a replication advantage by being associated with a fictional self. If so a selfplex will form.

One reason for making machines capable of imitation (apart from research purposes) would be to make them behave more like humans. This might be useful for some purposes, but they are more likely to be hybrid, imitation, digital-copying machines than simple imitating meme machines like us. In other words, they would acquire most of what they know by downloading it directly. If humans are also provided with memory chips, implanted additional processing capacity, and direct links to the Web, then we might all end up as such hybrids, as some people have suggested we will (Kurzweil, 1999). Assuming that the explosive increase in memes carries on then there will be intense memetic competition to get copied by these hybrid creatures and therefore pressure for selfplexes to form. Even such highly advanced creatures would fall for the illusion that they have an inner self who is conscious.

Finally, rather than thinking in terms of how 'we' should design future machines, it may be better to think in terms of how the memetic co-evolutionary process will design them. As machines copy more and more information from one to another, the machinery they use will co-evolve with that information, ultimately becoming self-replicating. We are not likely to have control over this process.

In the discussion so far, I assumed that one selfplex was linked with one physical machine, as it is with most human beings. There is no reason why this need always be so. Not only might there be the equivalent of multiple personalities within machines, but selfplexes might have no permanent physical home, being purely informational structures using whatever processing resources they can (Kurzweil, 1999). So long as there is competition to be copied by such entities, and the entities have boundaries that repel some memes and accept others, then selfplexes will form. Their illusions will be a little different from ours since they will not believe they inhabit or own or control a body, but they may still think that they are a conscious self having a stream of experiences.

Whether any of these machines will be able to transcend the illusion of consciousness remains a question for another time.

References

Aunger, R.A. (ed. 2000), *Darwinizing Culture: The Status of Memetics as a Science* (Oxford: OUP).
Baars, B.J. (1988), *A Cognitive Theory of Consciousness* (Cambridge: Cambridge University Press).
Batali, J. (1998), 'Computational simulations of the emergence of grammar', in *Approaches to the Evolution of Language: Social and Cognitive Bases*, ed. J.R. Hurford *et al.* (Cambridge: CUP).

Blackmore, S.J. (1998), 'Imitation and the definition of a meme', *Journal of Memetics: Evolutionary Models of Information Transmission*, **2**. http://www.cpm.mmu.ac.uk/jom-emit/1998/vol2/blackmore_s.html
Blackmore, S.J. (1999), *The Meme Machine* (Oxford: Oxford University Press).
Blackmore, S.J. (2000), 'Memes and the malign user illusion', Association for the Scientific Study of Consciousness Conference 'Unity and Dissociation', Brussels, July 2000.
Blackmore, S. (2001), 'Evolution and memes: The human brain as a selective imitation device', *Cybernetics and Systems*, **32**, pp. 225–55.
Blackmore, S.J. (2002), 'There is no stream of consciousness', *Journal of Consciousness Studies*, **9** (5–6), pp. 17–28.
Block, N. (1995), 'On a confusion about a function of consciousness', *Behavioral and Brain Sciences*, **18**, pp. 227–87.
Breazeal, C.L. (2001), *Designing Sociable Robots* (Cambridge, MA: MIT Press).
Brooks, R.A. (1991), 'Intelligence without representation', *Artificial Intelligence*, **47**, pp. 139–59 (also reprinted, with extra material, in Haugeland, ed. 1997, pp. 395–420).
Brooks, R.A., Breazeal, C., Marjanovic, M. Scassellati, B. and Williamson, M.M. (no date), 'The Cog project: Building a humanoid robot', http://www.ai.mit.edu/projects/cog/
Chalmers, D.J. (1995), 'Facing up to the problem of consciousness', *Journal of Consciousness Studies*, **2** (3), pp. 200–19.
Clark, A. (1997), *Being There: Putting brain, body, and world together again* (Cambridge, MA: MIT Press).
Dautenhahn, K. and Nehaniv, C.L. (ed. 2002), *Imitation in Animals and Artifacts* (Complex Adaptive Systems Series; Cambridge, MA: MIT Press).
Dawkins, R. (1976), *The Selfish Gene* (Oxford: Oxford University Press; new edition with additional material, 1989).
Deacon, T. (1997), *The Symbolic Species: The Co-evolution of Language and the Human Brain* (London: Penguin).
De Boer, B. (1997), 'Self-organisation in vowel systems through imitation', in *Proceedings of the Fourth European Conference on Artificial Life*, ed. P. Husbands and I. Harvey (Cambridge MA).
Dehaene, S. and Naccache, L. (2001), 'Towards a cognitive neuroscience of consciousness: basic evidence and a workspace framework', *Cognition*, **79**, pp. 1–37.
Dennett, D.C. (1991), *Consciousness Explained* (Boston and London: Little, Brown and Co.).
Dennett, D.C. (1995), *Darwin's Dangerous Idea* (London: Penguin).
Dennett, D.C (2001), 'Are we explaining consciousness yet?', *Cognition*, **79**, pp. 221–37
Donald, M. (1991), *Origins of the Modern Mind: Three Stages in the Evolution of Culture and Cognition* (Cambridge, MA: Harvard University Press).
Dunbar, R. (1996), *Grooming, Gossip and the Evolution of Language* (London: Faber & Faber).
Gallagher, S. and Shear, J. (ed. 1999), *Models of the Self* (Thorverton, Devon: Imprint Academic).
James,W. (1890), *The Principles of Psychology* (London: MacMillan).
Kirby, S. (2001), 'Spontaneous evolution of linguistic structure: An iterated learning model of the emergence of regularity and irregularity', *IEEE Transactions on Evolutionary Computation*, **5**, pp. 102–10
Kurzweil, R. (1999), *The Age of Spiritual Machines: How we will live, work and think in the new age of intelligent machines* (New York and London: Texere).
Matsuzawa, T. (ed. 2001), *Primate Origins of Human Cognition and Behavior* (Tokyo: Springer).
McGinn, C. (1987), 'Could a machine be conscious?', in *Mindwaves*, ed. C. Blakemore and S. Greenfield (Oxford: Blackwell).
McGinn, C. (1999), *The Mysterious Flame: Conscious Minds in a Material World* (New York: Basic Books).
Nagel,T. (1974), 'What is it like to be a bat?', *Philosophical Review*, **83**, pp. 435–50
Pickering, J. (ed. 1997), *The Authority of Experience: Essays on Buddhism and Psychology* (London: Curzon Press).
Pinker, S. (1994), *The Language Instinct* (New York: Morrow).
Pinker, S. (1997), *How the Mind Works* (London: Penguin).
Ridley, M. (2003), *Nature via Nurture: Genes, Experience and What Makes us Human* (London: Fourth Estate).
Searle, J. (1992), *The Rediscovery of the Mind* (Cambridge, MA: MIT Press).
Searle, J. (1997), *The Mystery of Consciousness* (New York: New York Review of Books).
Steels, L. (1998), 'The origins of syntax in visually grounded agents', *Artificial Intelligence*, **103**, pp. 1–24
Steels, L. (2000), 'Language as a complex adaptive system', in *Lecture Notes in Computer Science. Parallel Problem Solving from Nature* — PPSN-VI. Volume Eds. Schoenauer *et al.* (Berlin: Springer-Verlag).
Turing, A. (1950), 'Computing machinery and intelligence', *Mind*, **59**, pp. 433–60.
Varela, F.J. and Shear, J. (1999), *The view from within: First Person Approaches to the Study of Consciousness* (Thorverton, Devon: Imprint Academic).
Velmans, M. (2000), *Understanding Consciousness* (London and Philadelphia: Routledge).
Wills, C. (1993), *The Runaway Brain: The Evolution of Human Uniqueness* (New York: Basic Books).

Rodney M.J. Cotterill

CyberChild

A Simulation Test-Bed for Consciousness Studies

The first brief description is given of a project aimed at searching for the neural correlates of consciousness through computer simulation. The underlying model is based on the known circuitry of the mammalian nervous system, the neuronal groups of which are approximated as binary composite units. The simulated nervous system includes just two senses — hearing and touch — and it drives a set of muscles that serve vocalisation, feeding and bladder control. These functions were chosen because of their relevance to the earliest stages of human life, and the simulation has been given the name CyberChild. The system's pain receptors respond to a sufficiently low milk level in the stomach, if there is simultaneously a low level of blood sugar, and also to a full bladder and an unchanged diaper. It is believed that it may be possible to infer the presence of consciousness in the simulation through observations of CyberChild's behaviour, and from the monitoring of its ability to ontogenetically acquire novel reflexes. The author has suggested that this ability is the crucial evolutionary advantage of possessing consciousness. The project is still in its very early stages, and although no suggestion of consciousness has yet emerged, there appears to be no fundamental reason why consciousness could not ultimately develop and be observed.

Introduction

This paper is a preliminary report on a project aimed at searching for conscious behaviour in a modelled child. It is a contribution to a set of papers on machine consciousness. Some may feel that this collection is overly ambitious, given that science has not yet provided a definition of consciousness. And it is certainly true that this shortcoming hampers efforts to implement the phenomenon in a machine. But this should not deter us from constructing devices — actual or merely simulated — capable of displaying behaviour that would be taken as a sign of consciousness in a human being. The study being reported here has opted for simulation, because of the relative ease with which the model's parameters can be changed. It was decided from the outset that the modelled system should aim at simulating a rudimentary infant, and it was for this reason that the project was given the name CyberChild.

There have, of course, been numerous attempts to mimic the human mind with machines and computer programs. Before the Second World War, Turing (1936; 1950) had famously argued that any symbolic manipulation can be performed by a machine possessing a suitable program. McCulloch and Pitts (1943; see also Pitts and McCulloch, 1947, and McCulloch, 1965) were responsible for the early attempts to capture such manipulation in circuits based on nerve cells, their work initially focussing on the more primitive of the two routes in the visual system, namely that which joins the retina to the superior colliculus. Shannon and Weaver (1949) addressed the more general issue of how information progresses through a system, and the concurrent development of the stored-program electronic computer (invented by Atanasoff in the mid-1930s — see Mackintosh, 1988) led to speculations about the nervous system resembling such a device. This, in turn, gave rise to the branch of science known as *cybernetics* (Wiener, 1948; Ashby, 1952; von Neumann, 1958; Gardner, 1959). The development of generalised input–output devices and programs has continued (some of the earlier efforts being due to Taylor, 1959; Steinbuch, 1961; Rosenblatt, 1962; Minsky, 1969; Minsky and Papert, 1969; Aleksander and Burnett, 1987; Amit, 1989).

None of these projects led to systems for which their designers claimed the presence of consciousness. Indeed, most of the projects had run their course before consciousness came back into vogue as a research endeavour (it having received much attention around the turn of the previous century, not the least from James [1890], before it ran foul of Behaviourism a couple of decades later). Their constructors were understandably influenced by the wave of optimism generated by the advent of electronic computers, and the majority of them were perceived as contributions to the quest for artificial intelligence. This too was fraught with difficulties, given that there was no consensus as to how intelligence should be defined — or reliably measured.

The present project takes as its starting point the author's own theories of consciousness and intelligence (Cotterill, 1995; 1996; 1997a,b; 1998; 2001). They have always maintained that the internal feedback made possible by the efference-copy routes is a vital feature of the neural processes underlying consciousness. And they have been similarly resolute in identifying the premotor and supplementary motor regions (hereinafter PMA/SMA) of the cerebral cortex as the key sources of that internal feedback. The core circuitry thus involves loops, other important participants being the cingulate gyrus, the thalamic intralaminar nuclei, the nucleus reticularis thalami and of course the lobes that receive sensory input. These components were naturally included in the simulated system.

It seems rather unlikely, however, that the presence of consciousness in a simulation could be detected solely on the basis of observed correlation between certain neural signals. The project is therefore striving to represent reliably the interaction of the nervous system with the rest of the body, and through this with the body's surroundings. The modelled system included rudimentary muscles, such as those serving the voice and a primitive limb; a stomach and a bladder; receptors for various types of pain; and also receptors for touch and for hearing. The spindles associated with the muscles are also being included in the simulation.

The simulated child has its own drives, these being directed toward acquiring sufficient sustenance and avoiding discomfort. It has limited capacity for helping itself in these respects, for example by raising a feeding-bottle to its mouth when such a receptacle is placed in its touch-sensitive hands, and by controlling its urination by tensing its bladder muscle. But ultimately — because it cannot lift a bottle if no bottle is provided, and because it cannot indefinitely withhold urination — its survival and wellbeing depend upon its ability to communicate its needs to its human environment. And its behaviour is being monitored for signs that it is able to modify its repertoire of such communications. A close watch is also being kept to see whether it can vary its responses to sensory input.

It is felt that such modification and variation will be the most reliable indication that consciousness *could* be present in the simulation (Cotterill, 2001). No-one researching in this field would be impressed by an animated figure that verbally claimed that it was conscious; it would be a simple matter to program a simulated person to do *that*. Indeed, there is a classic philosophical problem connected with proving unequivocally the presence of consciousness in one's fellow human beings, despite any assurances they might give that they are indeed conscious. But we nevertheless do not doubt that they *are* conscious, and that they are not merely unthinking automatons — what Kirk (1974; see also Moody, 1994) has referred to as *zombies*. This conclusion stems from our observations of other people, and is based on a logic that invokes congruence; we are convinced of our own consciousness, and because other people resemble us, they too must be conscious. In the present project, similarly, one is looking for what could broadly be termed human-like behaviour.

Consciousness has not yet been detected in the model — indeed, there has not yet been the merest suggestion of the phenomenon — but the overall impression from the preliminary studies suggests that there is no obvious reason why this should not be possible.

The Model

The basic approach being used in the project is one of computing the state of the simulated child as a function of time. This is thus a process of iteration, in which the computational time step is as brief as the computer's central processing unit will permit. The time step will gradually be reduced, therefore, as the available computer power increases.

It must be emphasised that there is no attempt to link the computational time step to the duration of a nerve impulse (that is to say to the action potential). Thus although the simulated child's actions are driven by modelled neuronal units, it is the averaged state of each unit over the most recent time step that is important, rather than its possible firing behaviour.

The simulated child can consume simulated milk, if this is made available to it, and such consumption influences the glucose level in the child's blood. Care was taken to match these changes of blood glucose level with what would be expected from the energy equivalent of the amount of milk being consumed. Similarly, care was taken in relating the rate of fall of the blood glucose level to the rate of energy expenditure. The simulated child's time scale could be varied, so as to make things happen at a convenient rate, but the rates of energy intake and expenditure were then adjusted accordingly.

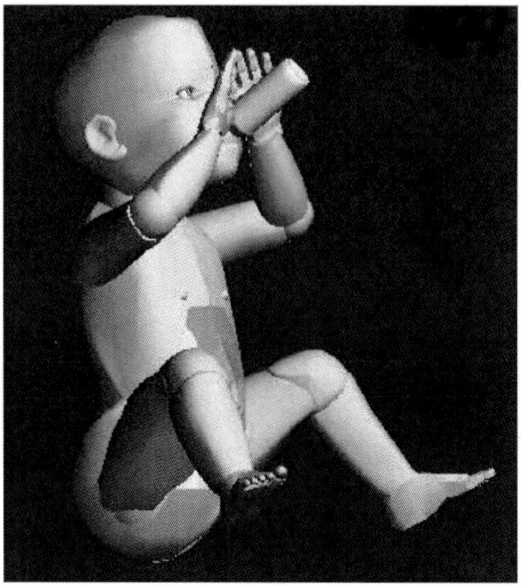

Figure 1. CyberChild Animation
'The child's actions can be continuously viewed on the computer screen'

The child is deemed to have expired if its blood glucose level drops to zero. Its means of avoiding such demise lies in its being able to utter sounds, which are the equivalent of crying. The experimenter reacts to the crying by providing milk in a simulated bottle, which the child can raise to its mouth with its simulated arm muscle. The child's actions can be continuously viewed on the computer screen (Figure 1), and this animation can be observed simultaneously with the current state of the simulated neuronal activities, as well as with the current state of such parameters as blood sugar level, stomach milk level, bladder urine level, vocal intensity level, and so on.

The consumed milk is gradually metabolised, thereby raising the blood glucose level, and the remnant liquid is converted to simulated urine, which accumulates in the bladder. Depending on the moment by moment tension of the simulated bladder muscle, this urine flows out into the simulated nappy (diaper) at a variable rate. Indeed, the flow is zero if the tension of the bladder muscle exceeds a certain threshold value. Putting it colloquially, the child could then be said to be holding its water. Conversely, when the amount of urine deposited in the nappy is sufficiently large, the bottom pain receptors start to dispatch nerve signals to the appropriate parts of the nervous system. If the pain exceeds a certain level, the simulated child cries. The experimenter can then respond by changing the nappy.

The various muscular components were represented by nominal variables. The simulated voice, for example, could not faithfully reproduce the full human vocal spectrum, because of limitations on the size of the computer model. The plan is ultimately to achieve the full spectrum, however, and to analyse external vocal input with fast Fourier transform software. This will make possible two-way voice communication with the simulated child.

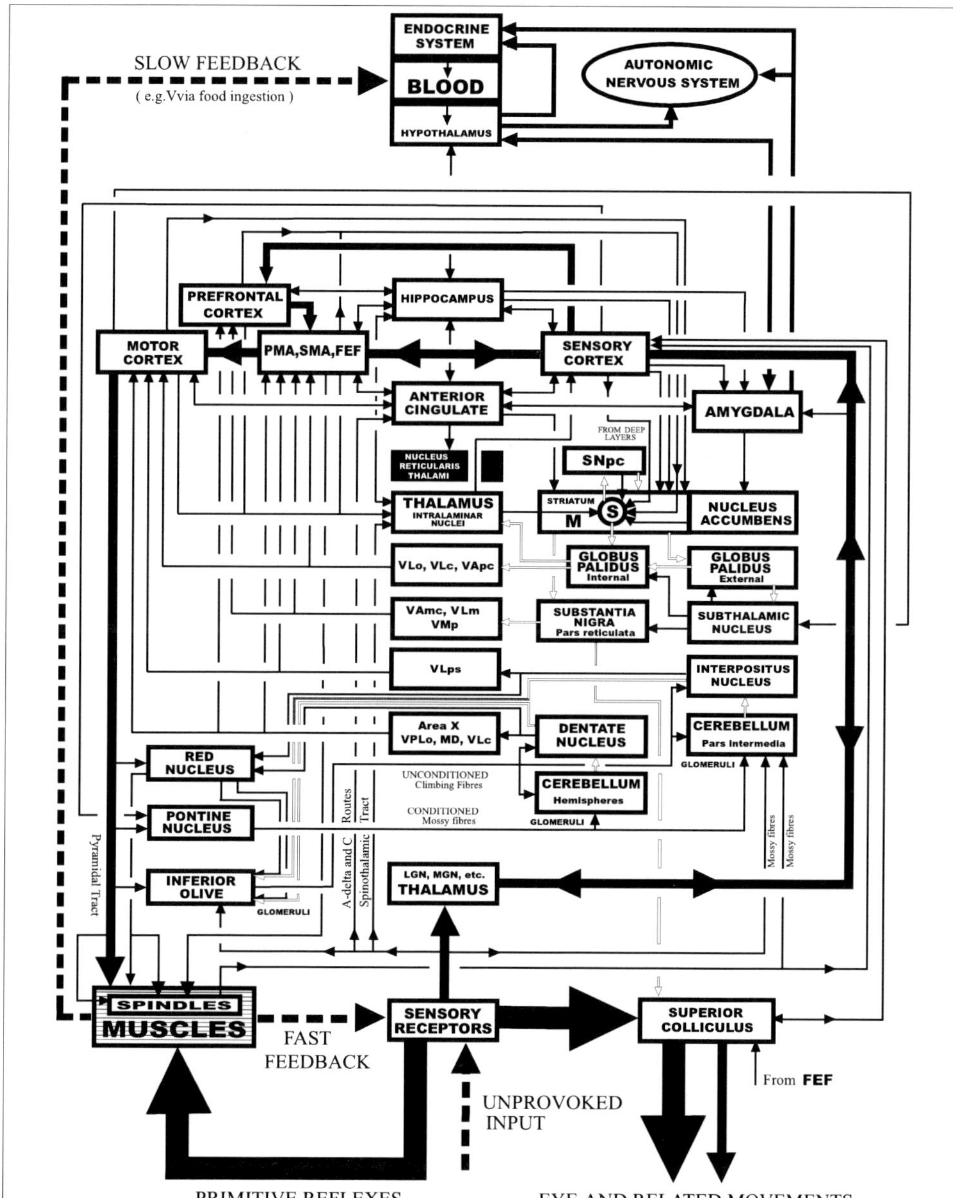

Figure 2. The circuit diagram used as the basis of the CyberChild simulation

The black and white lines represent excitatory and inhibitory routes, respectively, not individual axons. Similarly, branching of lines indicates duality of routes, not the presence of axon collaterals. The somewhat thicker black lines at the upper right indicate influence in a more general manner. Only two representative regions of the cerebellum are shown, to avoid unnecessary complexity. The abbreviations are: PMA, premotor cortex; SMA, supplementary motor cortex; FEF, frontal eye field; SNpc, substantia nigra pars compacta. The abbreviations for the thalamic nuclei are: LGN, lateral geniculate nucleus; MGN, medial geniculate nucleus; MD, medialis dorsalis; VAmc, ventralis anterior medialis caudalis; VApc, ventralis anterior pars principalis; VLc, ventralis lateralis caudalis; VLm, ventralis lateralis pars medialis; VLo, ventralis lateralis pars oralis; VLps, ventralis lateralis pars postrema; VMp ventralis medialis postralis; VPLo, ventralis posterior lateralis pars oralis.

The circuit

The central structure of the model is naturally the underlying neural circuit, because it is this which enables the various body components to interact. The present author has published a particularly detailed circuit for a typical mammal (Cotterill, 2001), and this is reproduced here as Figure 2. The caption to this figure lists the various abbreviations that have been used. Some features appearing in the circuit have not yet been incorporated in the simulation, this being the case for the endocrine system, for example.

A major decision in the design of the simulated model concerned the number of neuronal groups — that is to say minicolumns and ganglia — to be included in each of the brain regions depicted in Figure 2. A large number of these units would naturally increase the number of different possible patterns of neural activity, but it would tend to slow down the rate of computation. Conversely, the speed achievable with a smaller number of such units would be offset by possible limitations in the model's cognitive capacity.

It was ultimately decided to have 20 units in each brain component, and to strive for an average activity level in which about half of the units are active at any one time. The number of different possible patterns, when 10 units are active and 10 inactive, is of course simply the binomial coefficient $\binom{20}{10}$, the numerical value of which is about 10^5. It was hoped that this would provide the model with a sufficiently varied repertoire of responses, as long as one does not attempt to give the simulated child the full human vocal spectrum, as mentioned above.

Some of the brain components indicated in Figure 2 require special treatment. This is the case for the hippocampus, for example, which has 20 units in each of its component regions, that is to say the subiculum, the dentate gyrus, and the CA3 and CA1 regions. Similarly, the cerebellum needs to be able to capture numerous possible combinations of input patterns, so it too contains units intended to represent such different neural groups as the granule, stellate, basket, Golgi and Pukinje cells.

Efference copy and attention

Certain anatomical features shown in Figure 2 are not widely known among the artificial consciousness fraternity, and they thus deserve further comment. Moreover, while it is generally appreciated that the cortical hierarchy is not unidirectional (Felleman and Van Essen, 1991), there is less familiarity with the proposal that learning can take place in the reverse (that is to say backward-directed) cortical projections (Cotterill and Nielsen, 1991; Singer 1995). That latter feature of the brain's anatomy was incorporated in the present model. But it must be emphasised that such backward-directed routes are *not* the only conduits for signals travelling from the front of the brain to the back. There are also the two routes that link the PMA/SMA cortices (see PMA and SMA in Figure 2) with the sensory receiving lobes in the brain's posterior. One of these routes passes through the cingulate gyrus while the other passes through the thalamic intralaminar nuclei. *Both* of these routes are said to convey *efference-copy signals* from the PMA/SMA cortices to the back of the brain. The word *copy* is used because the signals are transported along routes that are collateral to the main routes from PMA/SMA to the primary motor cortex (M1).

The signals travelling along that latter route are clearly the ones that lead to muscular activation (if the signals are sufficiently strong, that is), whereas the efference-copy signals cannot activate muscles because their targets are not muscles. Those efference-copy signals must nevertheless be *identical* to the ones dispatched along the main route to M1, however, because they emanate from the *same* neurons, at the *same* time. It is for this reason that the term proxied movement has been coined (Cotterill, 2001) to describe what happens when the signals dispatched from PMA/SMA are not sufficiently strong to activate the muscles; the signals simultaneously dispatched through the efference-copy routes may nevertheless be strong enough to mediate thought.

Let us now consider what the efference-copy signals actually achieve. We should start by considering the case in which muscle activation *does* take place. This will cause the body to send signals (formally called *exefferent signals*) into the surroundings, as when activation of the vocal apparatus causes us to emit sound. Some of those signals (formally called *reafferent signals*) will impinge upon the sensory receptors and give rise to signals in the brain's sensory receiving areas, as when our ears receive signals associated with our own voice. (The term *self-paced environmental probing* has been used for this mechanism in Cotterill, 2001). But those sensory receiving areas will also have received efference-copy signals associated with the same muscle-activating-and-sensory-receiving event, possibly a few milliseconds earlier because the internal route is probably shorter in purely neural terms. The term *internal reafference* was coined as an alternative to the term efference copy (Cotterill, 1995), to stress what the latter are actually doing for the system. It seems reasonable to guess that a memory trace of the outcome of the self-paced probing event would not be laid down in the absence of the signal reinforcing effect of the efferent copy (or internal reafferent) signals.

As noted earlier, signals emanating from PMA/SMA with insufficient strength to activate muscles may nevertheless be able to evoke memories of the outcomes of previous self-paced probing of the environment. And from what has just been stated about the learning mechanism, it is clear the memories invoked would automatically be the ones relevant to the prevailing situation. Moreover, because the efference-copy routes involve those parts of the brain's circuitry known to be involved in attention (Cotterill, 1997a,b; 2001), their carrying of signals would just as automatically put the system on the alert, in a manner whose relevance was guaranteed by the prior experience laid down in memory. The anterior cingulate, for example, is known to be the target of the pain-information-dispatching A-delta and C routes (see Figure 2). The upshot of all this is that a pattern of neural activation in PMA/SMA will automatically invoke the probable outcome of the related pattern of muscular activation — if that pattern has ever occurred before, and if it has led to the laying down of a memory.

The mechanism just sketched strictly applies only to events in which the self-paced environmental probing is overt, that is to say events in which the muscles are indeed moved. This is probably a faithful reflection of the situation in the very earliest part of life, as in the present study of CyberChild. But it is important to ask how this same mechanism could serve perception of things beyond the scope of the body's muscles. Consider the example of a child's first observation of a feeding bottle. Such observation will be apparently passive; the child has played no part in the bottle's being there, or in the sound it makes when shaken by the parent. But *exafferent* (as opposed to *reafferent*) signals will be impinging upon the visual and auditory receiving areas of the cortex, and these will be giving rise to onward-travelling signals that ultimately result in signals arriving at the PMA/SMA regions. This will provoke activity in those two areas, and that activity, in turn, will cause efferent copy signals to be dispatched to those same sensory receiving areas. The upshot will be signals travelling around the closed-loop circuits variously routed through the cingulate gyrus and the thalamic intralaminar nuclei. So despite the novelty of this seemingly passive observation, signals will be generated that circulate around the closed loops. The stimulus will not in general be sufficient to provoke actual muscular contraction, however; hence the term *proxied movement*. (A very loud sound or intense flash of light could indeed provoke movement, but that is merely because the infant possesses innate reflexes for immediately reacting to such stimuli.)

What are such circulating proxied-movement signals doing for the inexperienced infant? Probably very little, and perhaps essentially nothing. This is primarily because they have not been causally linked to anything related to the child's drives; they are without significance because they have no *meaning* (Searle, 1980). But the situation will be changed when the apparently passive setting up of those circulating signals becomes associated with something relevant for the child's body, and with something achievable by its muscular movements. This starts to happen when the bottle is grasped, raised to the mouth, and later relieves any pangs of hunger that the child might be experiencing. (As is well known, infants are in fact born with grasping and sucking reflexes, and this clearly gives them an evolutionary advantage. A mother's anxiety is justified if she finds that her new-born child is not grasping and sucking normally.)

The neuronal units

There is much evidence indicating that the brain's neurons do not function individually. The neurons in the cerebral cortex, for example, appear to be arranged in what are known as minicolumns (Mountcastle, 1979; Szentagothai, 1983; Fujita *et al.*, 1992). These have the approximate shape of cylinders, deployed at right angles to the cortical surface. Each such minicolumn is comprised of a few hundred neurons, distributed among the various cortical layers, there being both excitatory and inhibitory cells. It seems unlikely that all the excitatory neurons in a given minicolumn would be activated every time that structure receives activation from elsewhere in the nervous system. On the contrary, it appears to be sufficient that just a few of those cells actively participate in the processing provoked by that activation.

The inhibitory neurons in a given minicolumn are known to pass on their negative influence to neighbouring minicolumns. This tends to enhance the relative activity of the most strongly activated minicolumns, the upshot being the familiar winners-take-all situation (also sometimes referred to by the expression feed-the-fattest).

In the present model, it was assumed that the average strength of such lateral inhibition would be automatically adjusted by the system, so as to lead to a situation in which a reasonably constant fraction of the minicolumns were activated at any instant. And it was further assumed that this fraction would lie around a half. This would tend to maximise the number of different activation patterns available to the system, and make the most rational use of the system's resources.

Memory

Ever since the pioneering suggestion by Hebb (1949), it has been generally accepted that learning involves changes at the synaptic level, and the idea has of course received strong endorsement through the work of Kandel and his colleagues (Kandel, 1979; Kandel and Hawkins, 1992). Hebbian learning has long since become a standard feature of neural network models having associative capabilities, such as those described by Freeman and Skarda (1987), and by Kohonen (1989).

The Hebbian learning mechanism operates over a time period that is long compared with that involved in consciousness, however. It is clear that a model aiming at simulating consciousness must also possess short-term memory. Evidence was obtained as early as forty years ago (Petersen and Petersen, 1959) that its time span is a few seconds, and this finding has not since been seriously challenged. The present author has linked consciousness to the time-dependent trajectory that the animal can prescribe through its own muscular hyperspace (Cotterill, 1997b), that is to say the totality of the body's possible muscular movements. One could loosely call this navigation through muscular hyperspace. As with navigation in the more colloquial sense, the mechanism requires at least brief retention of the route recently taken. If there is no such memory, proper navigation becomes impossible. Similarly, consciousness would be impossible if the system's span of short-term memory collapsed to zero; it would then be impossible to relate cause to effect.

The simulation model therefore required two distinct forms of memory: Hebbian learning, to permit gradual acquisition of behavioural skills, and short-term memory to permit consciousness. The first of these was incorporated in a manner aimed at computational efficiency. The mechanism involves synaptic changes that are conditional on the activity levels of the participating units, and this might seem to require manipulation of Boolean variables in the corresponding computer program (that is to say the incorporation of IF statements, or their equivalents). The use of such variables in a program's inner loops could be expensive in terms of central-processor resources.

But it transpires that an algorithm can be derived which permits direct computation of the change in synaptic efficiency (sometimes referred to as synaptic strength or synaptic weight). Let the efficiency of the synapse between unit i and unit j be S_{ij}, and let the activity level (or voltage) on those units be V_i and V_j. The change in synaptic efficiency can then be written

$$\Delta S_{ij} = \alpha\left(V_i + V_j\right)\left\{V_i V_j - S_{ij}\left[V_i + V_j - V_i V_j\right]\right\} \quad (1)$$

in which α is the rate of change of efficiency (or learning rate), this naturally being an adjustable variable.

If V_i and V_j are both unity (such normalisation being a frequently employed convenience in computer simulations of neural networks), the right-hand side of equation 1 takes on the value $2\alpha(1-S_{ij})$, which always has a positive value because S_{ij} is normalised to lie between zero and unity. In other words, the correlation between two currently active units leads to an increase in the efficiency of the synapse through which they are linked.

If, on the other hand, one of those units is currently inactive, the right-hand side of equation 1 takes the value $-\alpha S_{ij}$, which is clearly negative. So a mismatch between the activities of the two units in question leads to a decrease in the corresponding synaptic efficiency. It should also be noted that this decrease has only half the numerical value of the increase used for the active–active correlation. This is necessary because a mismatch can occur in two different ways, active–inactive or inactive–active, whereas a match occurs only when there is an active–active correlation. This latter type of event is thus only half as frequent. In both cases, the rate of change of synaptic efficiency is most rapid when its current value lies around the middle of its range, whereas the rate falls away toward zero when either of the extremities is approached.

So much for the long-term changes in memory. Let us turn now to the issue of short-term memory. It is an interesting feature of the circuit shown in Figure 2 that some regions of the thalamus send signals both to the PMA/SMA and to the prefrontal cortex. One could speculate as to the rationale of such duplicated dispatch. It seems likely that its desirability has something to do with the fact that the firing patterns of neurons in the prefrontal cortex have longer persistence times (Fuster, 1989). Indeed, this would automatically imbue the circuit with what could be called inertia, with the prefrontal cortex being unable to keep up with the faster changes happening in the PMA/SMA regions.

This suggests a relatively straightforward way in which short-term memory could be introduced into the model. Consider the neuronal unit i in a region receiving nerve signals and another unit j in the region dispatching those signals. The activity level in the receiving unit at the epoch $t + \Delta t$, due to signals having been dispatched at epoch t, will be given by

$$V_i(t + \Delta t) = \phi V_i(t) + \psi \sum_{j \neq i} S_{ji} V_j(t) \quad (2)$$

where ϕ and ψ are parameters that determine the span of short-term memory. If ϕ is zero, the current state of the receiving unit is not influenced by its past state, however recent, and its activity level is thus determined only by those units that are sending it signals from the other region of the brain. If, on the other hand, ψ is zero, the receiving unit is in the unlikely situation of being influenced *only* by its past state, and thus not by any other units, despite the fact that the latter might be vigorously dispatching their signals.

The most likely scenario is that both ϕ and ψ will be non-zero, and the short-term memory will increase its temporal span the greater is ϕ, relative to ψ. So the longer persistence times found in the prefrontal cortex, compared with those seen in the PMA/SMA cortices (Fuster, 1989), can easily be captured in the model by employing a higher ratio ϕ/ψ in the prefrontal cortex. This was the strategy adopted in the present work, and the behaviour of the model was studied as a function of the ϕ/ψ ratio.

Preliminary Observations

CyberChild has been programmed with the drive to survive, and it expires if it fails to acquire adequate nourishment. But in its neonatal state it is unable to provide for itself. It is born with the ability to cry, however, and thus to communicate with its surroundings, albeit in a rather rudimentary fashion. If it is hungry, by dint of its blood sugar level being low and its stomach empty, it can signal its plight to the investigator, who can then supply it with a bottle of milk, which it can raise to its mouth by activating its arm muscles.

The investigator places a filled bottle in CyberChild's hand, and the arm moves upward toward the mouth, in a series of steps that are dictated by the PMA/SMA cortices. The correct sequencing of these steps is necessary, if the bottle is to reach the mouth, and the sequence can be learned because of the reinforcing factor of the reward. In other words, the relevant synaptic efficiencies are increased if the child succeeds in obtaining the milk.

Similar learning is in operation in the excretory system, in that the bladder is controlled by a muscle that is also under the direction of the motor areas. This can enable CyberChild to prevent urination, that process otherwise leading to pain if the investigator omits to change the nappy (diaper). Prolonged retention of urine puts a strain on the bladder, however, and this too causes pain. So CyberChild is faced with the problem of rationalising its bladder movements so as to maximise its comfort, even though such comfort clearly cannot continue indefinitely. At the current stage of the investigation, such optimal adjustment has not been achieved by CyberChild.

CyberChild's vocal output is also dictated by the motor areas, of course, and its crying is its only means of soliciting help. This is thus a faithful reproduction of the situation that applies in the case of a real human child. There is the possibility that CyberChild will differentiate its vocalisations so as to signal its various needs, that is to say feeding vis-à-vis the changing of its nappy. Such differentiation has not yet been observed, but there seems to be no obvious reason why it could not ultimately emerge. If it does, this would indicate the ability to explore the novel concatenations of elementary muscular movements that the present author believes to underlie creativity (Cotterill, 2001).

At the time the manuscript of the present paper was submitted for publication, CyberChild was not displaying the hoped-for variation in its behaviour. On the contrary, it was tending to become trapped in stereotyped movement patterns. Moreover, it was either learning too rapidly or not at all. In other words, it has not yet proved possible to adjust the learning rate such that there is variety in its reactions to stimuli. Observation of the temporal development of the activity in the closed-loop circuits shows that this rapidly locks into a fixed pattern. Assuming that the time that elapses between the dispatch of signals from the neuronal groups of one area and their arrival at neuronal groups in another area is equivalent to about 10 ms of real time, which seems reasonable, the overall pattern becomes entrained within about 30 ms. This is much shorter than the time usually envisioned for an elementary cognitive event, namely about 200 ms.

There is clear evidence that 200 ms is indeed the approximate time that must elapse if the PMA/SMA cortices are to have sufficient time to select an appropriate activity pattern. It is well known, for example, that the eyes make about four saccades per second (that is, about one movement every 250 ms or so) when we observe a scene, and Thompson *et al.* (1996) have studied the temporal evolution of firing frequencies of neurons in the macaque monkey frontal eye field, after presentation of either a target stimulus or a distractor. (The frontal eye field can be regarded as the ocular counterpart of PMA/SMA.) The responses initially followed the same time course, irrespective of whether it was the target or the distractor that was presented. But the response profiles diverged after about 100 ms, the reaction to the distractor then falling away toward zero, whle the reaction to the target remained essentially constant. The separation of the target response and the distractor response was at its maximum around 200 ms.

There is a tendency for patterns of neural signals to persist for a certain period of time in what are known as *chaotic attractors* (Freeman and Skarda, 1987). The 200 ms just discussed is probably the time required for the signals to self-organise, as they circulate around the closed neuronal loops invoked above in the proposed consciousness mechanism. Such attractors are said to be *structurally stable* in that they resist small variations in their underlying dynamics. Because of the existence of an underlying attractor, the dynamics of the neural signals will be characterised by stable limit cycle organisation. It is possible that the long persistence time observed by Fuster (1985; 1997) arises *because* of temporary entrapment of signals in such a limit cycle.

The large difference between the entrapment time currently being observed in CyberChild's brain circuit and its counterpart in the real brain may be attributable to the great difference in the numbers of neurons in the relevant areas of the real brain and the numbers of neuronal units in CyberChild's brain; one would expect the entrapment time to be shorter if there are fewer units. This could be compensated for by suitable adjustment of the ϕ/ψ ratio in equation 2 above, but such adjustment might obscure a more profound malaise stemming from the small number of CyberChild's computational units; the number of possible activity patterns in the PMA and SMA might simply be insufficient.

There is a related issue which concerns the reverse projections discussed earlier. Zeki and Shipp (1988) discovered that the angular spread of these counter-running axons is considerably greater than that of their forward-directed counterparts, and it has been suggested that the consequently greater correlation- capturing capacity of the reverse projections could have a bearing on creativity (Cotterill, 2001). In the current version of CyberChild's brain circuit, the number of units in the various areas is simply too small to permit incorporation of the directional imbalance found by Zeki and Shipp. So here too, we have a possible source of the present simulation's tendency to get stuck in a single behavioural mode.

Happily, the computational power of computers continues to increase, both regarding working memory size and frequency of basic operations. As CyberChild's circuitry expands, some of the difficulties now being experienced may gradually disappear.

Discussion

CyberChild could be called an extreme example of the bottom-up approach. An actual simulation of the type described here can be carried out irrespective of whether one believes that it has any chance of displaying what one might conclude was conscious behaviour. Units representing those in the experimentally determined neural circuit are merely set up in algorithmic form, and their interactions are studied by computational techniques that are reasonably routine. The simulation is in this respect the antithesis of the top-down approach epitomised by such commendable efforts as those of Simon (1989) and Franklin (1995). In those studies, there is the implicit assumption that one knows what consciousness and intelligence are, and one then attempts to embody those attributes in a suitable computer program. The emphasis is on producing software with useful capabilities. The bottom-up approach, on the other hand, seeks to mimic the underlying biology as closely as possible, and the investigator hopes that interesting behaviour *could* emerge, if the model is sufficiently realistic.

The present investigator feels that working with such a simulation is a salutary experience. There is naturally the implicit assumption that consciousness requires no mystical ingredient, beyond what is comprised in the simulated circuit. But it takes a lot of faith in the reductionist canon to believe that consciousness will one day emerge from the blinking lights that are the simulation's graphical representations of the neuronal units (that is to say their on-screen depiction). Working with such a model forces one to face issues that might otherwise be glossed over. The question of drive has to be considered, for example. CyberChild is born with a set of priorities, of course, but what is the mechanism whereby these drives could become more nuanced and sophisticated?

Above all, direct interaction with such a simulation (and its associated animation) leads one to appreciate the profundity of Searle's speculations about the importance of meaning (Searle, 1980). In CyberChild, one sees the nervous system pared down to its essentials, and the issue of meaning is seen at its starkest. If evidence of conscious behaviour does emerge, therefore, one could reasonably hope that its neural correlates would be detectable, even though that is far from being a foregone conclusion. If such emergence and detection do materialise, we may have taken a significant step toward realising the long-cherished dream of creating *Homo siliciens*: consciousness in a computer.

Acknowledgements

It is a pleasure to acknowledge the continuing support of the Danish Natural Sciences Research Council. Warm thanks are also due to Kris Kegel and Søren Steno, whose collaboration with the author on animations (as opposed to simulations) of CyberChild will be published separately. The basic approach reported on here is covered by the patent EP(UK)0877984.

References

Aleksander, I. and Burnett, P. (1987), *Thinking Machines: The Search for Artificial Intelligence* (Oxford: Oxford University Press).

Amit, D.J. (1989), *Modelling Brain Function: The World of Attractor Neural Networks* (Cambridge: Cambridge University Press).

Ashby, W.R. (1952), *Design for a Brain* (London: Chapman and Hall).

Cotterill, R.M.J. (1995), 'On the unity of conscious experience', *Journal of Consciousness Studies,* **2** (4), pp. 290–312.

Cotterill, R.M.J. (1996), 'Prediction and internal feedback in conscious perception', *Journal of Consciousness Studies,* **3** (3), pp. 245–66.

Cotterill, R.M.J. (1997a), 'On the mechanism of consciousness', *Journmnal of Consciousness Studies*, **4** (3), pp. 231–47.

Cotterill, R.M.J. (1997b), 'Navigation, consciousness and the body/mind "problem"', *Psyke & Logos*, **18**, pp. 337–41.

Cotterill, R.M.J. (1998), *Enchanted Looms* (Cambridge: Cambridge University Press).

Cotterill, R.M.J. (2001), 'Cooperation of the basal ganglia, cerebellum, sensory cerebrum and hippocampus: Possible implications for cognition, consciousness, intelligence and creativity', *Prog. Neurobiol.*, **64** (1), pp. 1–33.

Cotterill, R.M.J. and Nielsen, C. (1991), 'A model for cortical 40 Hz oscillations invokes inter-area interactions', *NeuroReport*, **2**, pp. 289–92.

Felleman, D.J. and Van Essen, D.C. (1991). 'Distributed hierarchical processing in the primate cerebral cortex', *Cerebral Cortex*, **1**, pp. 1–47.

Franklin, S. (1995), *Artificial Minds* (Cambridge, MA: MIT Press).

Freeman, W.J. and Skarda, C.A. (1987), 'How brains make chaos in order to make sense of the world', *Behavioral and Brain Sciences*, **10**, pp. 161–95.

Fujita, I., Tanaka, K., Ito, M. and Cheng, K. (1992), 'Columns for visual features of objects in monkey inferotemporal cortex', *Nature*, **360**, pp. 343–6.

Fuster, J.M. (1985), *The Prefrontal Cortex: Anatomy, Physiology, and Neuropsychology of the Frontal Lobe* (New York: Raven Press).

Fuster, J.M. (1997), 'The prefrontal cortex and the temporal organization of behavior', in *The Association Cortex: Structure and Function*, ed. H. Sakata, A. Mikami and J.M. Fuster (Singapore: Harwood).

Gardner, M. (1959), *Logic Machines and Diagrams* (New York: McGraw-Hill).

Hebb, D.O. (1949), *The Organization of Behaviour: A Neuropsychological Theory* (New York: Wiley).

James, W. (1890), *The Principles of Psychology* (New York: Henry Holt).

Kandel, E.R. (1979), 'Small systems of neurons', *Scientific American*, **241** (3), pp. 66–76.

Kandel, E.R. and Hawkins, R.D. (1992), 'The biological basis of learning and individuality', *Scientific American*, **267** (3), pp. 52–60.

Kirk, R. (1974), 'Sentience and behaviour', *Mind,* **83**, pp. 60–1.

Moody, T.C. (1994), 'Conversations with zombies', *Journal of Consciousness Studies*, **1** (2), pp. 196–200.

Kohonen, T. (1989), *Self-Organization and Associative Memory* (Berlin: Springer-Verlag).

Mackintosh, A.R. (1988), 'Dr. Atanasoff's computer', *Scientific American*, **259** (2), pp. 90–6.

McCulloch, W.S. (1965), *Embodiments of Mind* (Cambridge, MA: MIT Press).

McCulloch, W.S. and Pitts, W.H. (1943), 'A logical calculus of the ideas imminent in nervous activity', *Bulletin of Mathematical Biophysics*, **5**, pp. 115–33.

Minsky, M. (1969), *Semantic Information Processing* (Cambridge, MA: MIT Press).

Minsky, M. and Papert, S. (1969), *Perceptrons* (Cambridge, MA: MIT Press).

Mountcastle, V.B. (1979), 'An organizing principle for cerebral function: the unit module and the distributed system', in *Neurosciences Fourth Study Program,* ed. F.O. Schmitt (Cambridge, MA: MIT Press).

Petersen, L.R. and Petersen, M.J. (1959), 'Short-term retention of individual verbal items', *J. Experimental Psychology: Learning, Memory and Cognition*, **58**, pp. 193–8.

Pitts, W.H. and McCulloch, W.S. (1947), 'How we know universals: the perception of auditory and visual forms', *Bulletin of Mathematical Biophysics*, **9**, pp. 127–47.

Rosenblatt, F. (1962), *Principles of Neurodynamics: Perceptrons and the Theory of Brain Mechanisms* (Washington, DC: Spartan Books).

Searle, J. (1980), 'Minds, brains and programs', *Behavioral and Brain Sciences*, **3**, pp. 417–57.

Shannon, C.E. and Weaver, W. (1949), *The Mathematical Theory of Communication* (Urbana: University of Illinois Press).

Simon, H.A. (1989), *Models of Thought* (New Haven: Yale University Press).

Singer, W. (1995), 'Development and plasticity of cortical processing architectures', *Science*, **270**, pp.758–64.

Steinbuch, K. (1961), 'Die lernmatrix', *Kybernetik*, **1**, pp. 36–45.

Szentagothai, J. (1983), 'The modular architecture principle of neural centers', *Reviews of Physiology, Biochemistry and Pharmacology*, **98**, pp. 11–61.

Taylor, W.K. (1959), 'Pattern recognition by means of analogous automatic apparatus', *Proceedings of the Institute of Electrical Engineers, London*, **106B**, pp. 168–72.

Thompson, K.G., Hanes, D.P., Bichot, N.P. and Schall, J.D. (1996), 'Perceptual and motor processing stages identified in the activity of macaque frontal eye field neurons during visual search', *Journal of Neurophysiology*, **76**, pp. 4040–55.

Turing, A.M. (1936), 'On computable numbers with an application to the Entscheidungs-problem', *Proceedings of the London Mathematical Society, Series 2*, **42**, pp. 230–67.

Turing, A.M. (1950), 'Computing machinery and intelligence', *Mind*, **59**, pp. 433–60.

von Neumann, J. (1958), *The Computer and the Brain* (New Haven: Yale University Press).

Wiener, N. (1948), *Cybernetics: or Control and Communication in the Animal and the Machine* (Cambridge, MA: MIT Press).

Zeki, S.M. and Shipp. S. (1988). 'The functional logic of cortical connections', *Nature*, **335**, pp. 311–17.

Stan Franklin

IDA

A Conscious Artifact?

After discussing various types of consciousness, several approaches to machine consciousness, software agent, and global workspace theory, we describe a software agent, IDA, that is 'conscious' in the sense of implementing that theory of consciousness. IDA perceives, remembers, deliberates, negotiates, and selects actions, sometimes 'consciously'. She uses a variety of mechanisms, each of which is briefly described. It's tempting to think of her as a conscious artifact. Is such a view in any way justified? The remainder of the paper considers this question.

Consciousness

The term 'consciousness' is used with so many different meanings and connotations that at least one philosopher/computer scientist is reluctant to use the term at all for fear of equivocation (Aaron Sloman, personal communication). In this section we'll explore some of those meanings in the hope of making clear the usage of the term in the following sections. In ordinary, day to day language the various meanings are distinguished by context, more or less successfully.

Perhaps the most common usage of the term refers to *phenomenal consciousness*, sometimes called sentience, or subjective (first-person) experience. Like trying to explain the colour purple to a person who has been sightless from birth, phenomenal consciousness is impossible to convey in words, but must be experienced to be understood. The term 'conscious' as used in the title is intended to convey phenomenal consciousness.

But phenomenal consciousness comes in several varieties, which are experienced differently. There are the altered states of consciousness produced by various hallucinogenic drugs. There is the phenomenal consciousness of dreams. There are the in-between states sometimes experienced while going to sleep or when waking up. There's the phenomenal consciousness of an infant as yet mostly uninfluenced by language. There's the phenomenally conscious state of mental quiet achieved by meditators who can turn off their inner voices and experience the world without the usual running verbal commentary. There is the phenomenal consciousness of my cat who certainly experiences things differently from me if only by virtue of the differences in our visual systems. There's the phenomenal consciousness of a snake that is incapable of fusing information derived from its various senses. And what about the phenomenal consciousness of a frog, likely so primitive? Does a mosquito have any at all? If not how does it seek heat and avoid certain smells? What about an ant following a pheromone trail? Many people want to attribute phenomenal consciousness only to humans, others to those organisms possessing neuropil (Freeman, 1995), and yet others only to biological organisms. A recent paper argues for phenomenal consciousness in all mammals (Baars, 2001).

Consciousness has many functions. It helps us deal with novel or problematic situations for which we have no automatized response. It makes us aware of potentially dangerous situations. It alerts us to opportunities presented by the environment. It allows us to perform tasks that require knowledge of location, shape, size or other features of objects in our environments. And there are a number of other functions of consciousness. I say that an agent (see below) possesses *functional consciousness* if its architecture and mechanisms allow it a number of these and, perhaps, other functions. (Functional consciousness is an example of a cluster concept, that is one that can be ascribed to something on the basis of it satisfying some sufficient but unspecified subset of a given collection of features [Sloman, 1995]. Surprisingly such concepts, though not well defined, turn out to be useful.) The 'consciousness' attributed in this article to IDA and other software agents is meant to be functional consciousness, and the quotes are meant to so indicate. Consciousness attributed to humans must be distinguished as to type by the context.

Can an agent (organism or otherwise) have functional consciousness without phenomenal consciousness? This question has vexed philosophers for decades (Chalmers, 1996). They refer to such a humanoid agent as a 'zombie'. Are there zombies? Could there be? We'll return to a version of this question in the next section.

Another type of consciousness that's often spoken of is self-consciousness, that is being aware of oneself. This often includes one's self image. Presumably, self-consciousness presupposes phenomenal consciousness. Damasio has recently described self-consciousness as beginning with the proto-self, that is, unconscious knowledge about the current state of the body (Damasio, 1999). The question of self-consciousness in primates, particularly in the other apes has been studied using a 'does he recognize himself in a mirror test' (Gallup, 1982). The latest results seem to indicate that all apes are self-conscious, and that no other primate is.

There's recently been a great deal of interest in possible ways of ascertaining or measuring phenomenal consciousness. One means that has been suggested is reportability (Baars, 1988). Being able to report, verbally or otherwise, the contents of one's awareness is often given as evidence of phenomenal consciousness, particularly when the reports can be confirmed by an experimenter. We'll discuss reportability again below.

Another approach is via neural correlates of consciousness as measured by PET or fMRI scans. If it were found that activity in a particular neural structure correlated well with phenomenal consciousness, the extent of such activity could be used as a measure independent of reportability. To my knowledge such correlates have not been found. To the contrary, several researchers hypothesize that activity over large areas, particularly cortical and thalamic, is required for consciousness, both phenomenal and functional (Freeman, 1999; Edelman & Tononi, 2000; Baars, 2002).

Machine Consciousness

What about the possibility of machine consciousness? Is it possible for a robot or a software agent (see below) to be functionally conscious? Phenomenally conscious? The answer to one question seems clearly yes. I will argue below that IDA is functionally conscious. The question of machine phenomenal consciousness seems much more difficult. The philosophers have engaged in considerable debate over this issue (Chalmers, 1996). In May of 2000 a workshop attended by two dozen leading researchers, psychologists, neuroscientists, philosophers, and AI scientists, were unable to make any noticeable headway with the problem.

There are several serious, ongoing projects aimed a producing machine consciousness. One such, headed by Igor Aleksander, is relatively far along, having produced a working system, MAGNUS, using neural modelling that's arguably capable of imagination (Aleksander, 2000). A second such is my own IDA project to be described in some detail below. IDA is currently up and running, and exhibiting functional consciousness. A third such project, conceived by Rodney Cotterill and also based on neural modelling, aims at developing machine consciousness in a manner analogous to the way a human child develops. An early version of 'CyberChild' is being demonstrated (Cotterill, 2001, and in this volume). For a fourth, Owen Holland and Rodney Goodman have embarked on a bottom up approach of adding additional capabilities to a robotic system until it shows signs of consciousness REF (see this volume). A fifth, due to Lee McCauley (2002), builds consciousness into a neural schema system REF.

A major question confronting the builders of all these systems is how to determine if, or when, their system has become phenomenally conscious. Though there seems to be no good answer, we'll return to this critical question later.

Global Workspace Theory

The material in this section is from Baars' two books (1988; 1997) and superficially describes his global workspace (GW) theory of consciousness.

In his global workspace theory, Baars, along with many others (e.g. Minsky, 1985; Ornstein, 1986; Edelman, 1987), postulates that human cognition is implemented by a multitude of relatively small, special purpose processes, almost always unconscious. (It's a multi-agent system.) Communication between them is rare and over a narrow bandwidth. Coalitions of such processes find their way into a global workspace (and into consciousness). This limited capacity workspace serves to broadcast the message of the coalition to all the unconscious processors, in order to recruit other processors to join in handling the current novel situation, or in solving the current problem. Thus consciousness in this theory allows us to deal with novel or problematic situations that can't be dealt with efficiently, or at all, by habituated unconscious processes. In particular, it provides access to appropriately useful resources, thereby solving the relevance problem.

This theory offers an explanation for consciousness being serial in nature rather than parallel as is common in the rest of the nervous system. Messages broadcast in parallel would tend to overwrite one another making understanding difficult. It similarly explains the limited capacity of consciousness as opposed to the huge capacity typical of long-term memory and other parts of the nervous system. Large messages would be overwhelming to small, special-purpose processors.

All this activity of processors takes place under the auspices of contexts (see Figure 1): goal contexts, perceptual contexts, conceptual contexts, and/or cultural contexts. Baars uses goal hierarchies, dominant goal contexts, a dominant goal hierarchy, dominant context hierarchies, and lower level context hierarchies. Each context is, itself, a coalition of processes. Though contexts are typically unconscious, they strongly influence conscious processes. A key insight of global workspace says that each context is, in fact, a coalition of processors.

Long back, William James proposed the ideomotor theory of voluntary action (James, 1890). James suggests that any unopposed idea that pops to mind is accepted and acted upon. He speaks at length of the case of deciding to get out of a warm bed into an unheated room in the dead of winter. 'This case seems to me

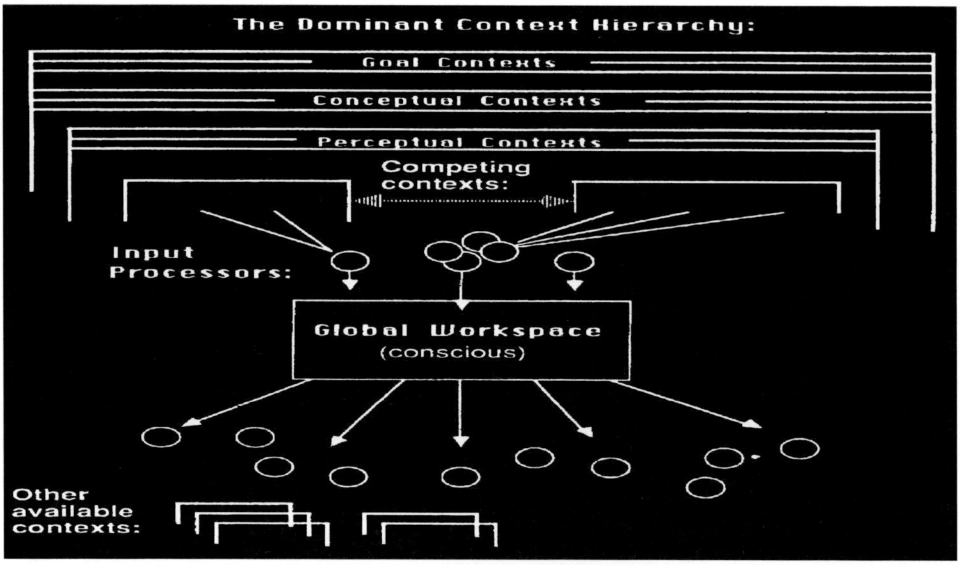

Figure 1. Global Workspace Theory

to contain in miniature form the data for an entire psychology of volition.' Global workspace theory adopts James' ideomotor theory as is, and provides a functional architecture for it (Baars, 1997 , Chapter 6).

Baars postulates that learning results simply from conscious attention, that is, that consciousness is sufficient for learning. There's much more to GW theory, including attention, action selection, emotion, voluntary action, metacognition and a sense of self. I think of it as a high level theory of cognition.

Software Agents

Artificial intelligence pursues the twin goals of understanding human intelligence and of producing intelligent software and/or artifacts. Designing, implementing and experimenting with autonomous agents furthers both these goals in a synergistic way. An *autonomous* agent (Franklin & Graesser, 1997) is a system situated in, and part of, an environment, which senses that environment, and acts on it, over time, in pursuit of its own agenda. In biological agents, this agenda arises from evolved drives and their associated goals; in artificial agents from drives and goals built in by its creator. Such drives, which act as motive generators (Sloman, 1987) must be present, whether explicitly represented, or expressed causally. The agent also acts in such a way as to possibly influence what it senses at a later time. In other words, it is structurally coupled to its environment (Maturana 1975; Maturana and Varela, 1980). Biological examples of autonomous agents include humans and most animals. Non-biological examples include some mobile robots, and various computational agents, including artificial life agents, software agents and many computer viruses. We'll be concerned with autonomous software agents, designed for specific tasks, and 'living' in real world computing systems such as operating systems, databases, or networks.

A *'conscious' software agent* is one that implements global workspace theory. The scare quotes are included to remind the reader that it's functional consciousness that's being claimed, not phenomenal consciousness. In addition to modelling this theory (Franklin & Graesser, 1999), such 'conscious' software agents should be capable of more adaptive, more human-like operations, including being capable of creative problem solving in the face of novel and unexpected situations.

Quick Overview of the IDA Model

IDA is one recent instantiation of our theory of intelligent software agents. IDA (Intelligent Distribution Agent) is a 'conscious' software agent that was developed for the US Navy (Franklin *et al.*, 1998). At the end of each sailor's tour of duty, the sailor is assigned to a new billet. This assignment process is called distribution. The Navy employs some 280 people, called detailers, to effect these new assignments. IDA's task is to facilitate this process by completely automating the role of detailer. IDA must communicate with sailors via email in natural language, by understanding the content and producing life-like responses.

Sometimes she will initiate conversations. She must access several databases, again understanding the content. She must see that the Navy's needs are satisfied by adhering to some ninety policies and seeing that job requirements are fulfilled. She must hold down moving costs, but also cater to the needs and desires of the sailor as well as is possible. This includes negotiating with the sailor via an email correspondence in natural language. Finally, she must write the orders and start them on the way to the sailor. At this writing an almost complete version of IDA is up and running and had been demonstrated to the satisfaction of the Navy.

In this article we will discuss IDA both as a conceptual model and as a computational model of global workspace theory. The conceptual model includes modules that have been designed, including mechanisms, but not implemented. Some unimplemented modules are important to our cognitive modelling but not to the Navy's application. Others, also not critical to the application, have been designed since the demonstration. In the subsections that follow, the modules of the conceptual model will be briefly described with references given to full descriptions. Unimplemented modules will be so indicated, as will the concordance between concepts in our model and those in global workspace theory. For science, the conceptual model is useful as a source of, hopefully testable, hypotheses (Franklin, 1997). Each design decision taken gives rise to the hypothesis that humans 'do it' that same way. The conceptual model, with its specified mechanisms is sufficient for this process of hypothesizing. Implementation is fine, but not required.

Developed during 1996–2001, the IDA computational model is the product of the 'Conscious' Software Research Group at the University of Memphis (http://csrg.cs.memphis.edu/CSRG/).[1] Built on a Java platform, the running IDA consists of approximately a quarter-million lines of code, and almost completely consumes the resources of a 2001 high-end workstation. Her various modules are briefly described below.

Following Hofstadter's terminology (see below) a codelet is a special purpose, relatively independent, mini-agent typically implemented as a small piece of code running as a separate thread. IDA depends heavily on such codelets in almost every module. This is as it should be since codelets in our model correspond with processors in global workspace theory, which are postulated to form the basis of human cognition. In what follows we will encounter several different types of codelets such as perceptual codelets, attention codelets, information codelets, behaviour codelets and language generation codelets. Many codelets play the role of demons (as in an operating system) waiting patiently for the conditions under which they can act. Some codelets subserve some higher-level construct, while others act completely independently.

The following sections contain brief descriptions of the various modules that make up the IDA architecture. The names of several of these modules, perception, emotion, deliberation, metacognition, consciousness, etc., are borrowed from folk psychology. This choice of terminology has the advantage of calling to

[1] More than twenty-five researchers contributed to its development. Their names can be found on the web site. (See the Acknowledgement section below for an abbreviated list.)

mind the functionality of the various modules. But there are tradeoffs. Sometimes the meanings are different in more or less subtle ways. For instance, perception to the psychologists includes conscious awareness, while IDA's perception module is preconscious. Also, IDA's consciousness module conforms functionally to global workspace theory, but whether it includes phenomenal consciousness is precisely the major question to be considered.

Perception

Perception in IDA consists mostly of processing incoming email messages in natural language (Zhang *et al.*, 1998b). In sufficiently narrow domains, natural language understanding may be achieved via an analysis of surface features without the use of a traditional symbolic parser (Jurafsky & Martin, 2000). Allen (1995) describes this approach to natural language understanding as complex, template-based matching. Ida's relatively limited domain requires her to deal with only a few dozen or so distinct message types, each with relatively predictable content. This allows for surface level natural language processing. We hypothesize that much of human language understanding results from a combined bottom up/top down passing of activation through a hierarchical conceptual net with the most abstract concepts in the middle. Thus IDA's language-processing module has been implemented as a Copycat-like architecture with perceptual codelets that are triggered by surface features and a slipnet (Hofstadter & Mitchell, 1994), a semantic net that passes activation. The slipnet

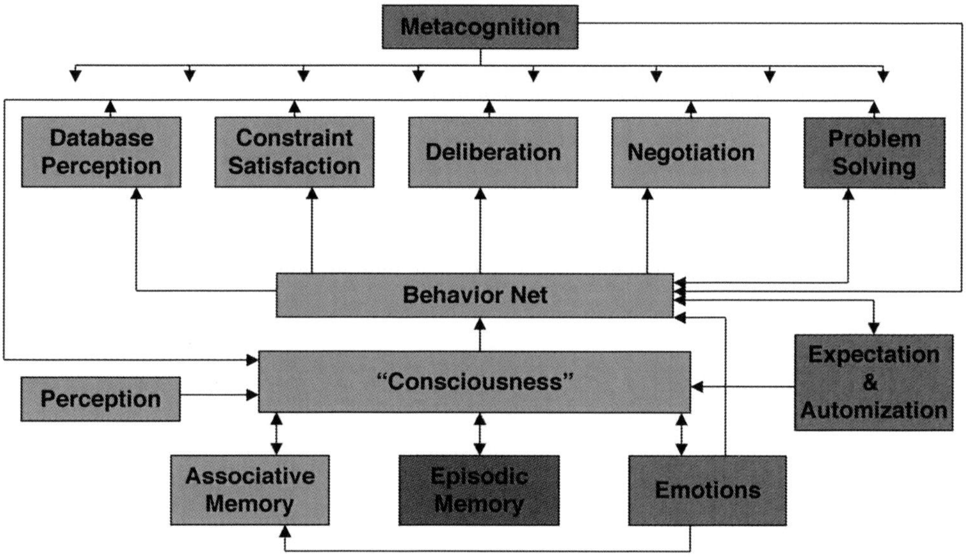

Figure 2. The IDA Architecture

stores domain knowledge. In addition there's a pool of perceptual codelets (processors) specialized for recognizing particular pieces of text, and production templates used by codelets for building and verifying understanding. Together they constitute an integrated sensing system for IDA, allowing her to recognize, categorize and understand.

For the purposes of this paper it's important to be clear about what is claimed by the word 'understand' as used in the previous sentence. An example may help. A secretary sending out an email announcement of an upcoming seminar on Compact Operators on Banach Spaces can be said to have understood the organizer's request that she do so even though she has no idea of what a Banach space is much less what compact operators on them are. In most cases it would likely require a person years of diligent effort to impart such knowledge. Nonetheless, the secretary understands the request at a level sufficient for her to get out the announcement. In the same way IDA understands incoming email messages well enough to do all the things she needs to with them. An expanded form of this argument can be found in my *Artificial Minds* (1995). Glenberg (1997) also makes a similar argument.

IDA must also perceive the contents read from databases, a much easier task. An underlying assumption motivates our design decisions about perception. Suppose, for example, that IDA receives a message from a sailor saying that his projected rotation date (PRD) is approaching and asking that a job be found for him. The perception module would recognize the sailor's name and social security number, and that the message is of the please-find-job type. This information would then be written to the workspace. The general principle here is that the contents of perception are written to working memory before becoming conscious.

Workspace

IDA solves routine problems with novel content. This novel content goes into her workspace, which is roughly plays the same role as human working memory. IDA's workspace is less dynamic, consisting of registers set aside for particular categories of information. Perceptual codelets write to the workspace as do other, more internal codelets. Quite a number of codelets, including attention codelet (see below) watch what's written in the workspace in order to react to it. IDA's workspace also contains registers dedicated to providing an interface with her long-term associative memory. (Modelling working memory as distinct from long-term memory puts us at odds with the best current thinking of cognitive scientists. We're working on the problem.) Items in the workspace decay over time, and may be overwritten. Not all of the contents of the workspace eventually make their way into consciousness.

Associative memory

IDA employs sparse distributed memory (SDM) as her major associative memory (Kanerva, 1988; Anwar *et al.* 1999; Anwar & Franklin, to appear). SDM is a content addressable memory that, in many ways, is an ideal computational mechanism for use as a long-term associative memory (LTM). Any item written to the workspace cues a retrieval from LTM, returning prior activity associated with the current entry. In our example, LTM will be accessed as soon as the message information reaches the workspace, and the retrieved associations will be also written to the workspace.

At a given moment IDA's workspace may contain, ready for use, a current entry from perception or elsewhere, prior entries in various states of decay, and associations instigated by the current entry, i.e. activated elements of LTM. IDA's workspace thus consists of both short-term working memory (STM) and something very similar to the long-term working memory (LT-WM) of Ericsson and Kintsch (1995).

Part, but not all, of the workspace, called the *focus*[2] by Kanerva (1988), is set aside as an interface with long-term LTM. Retrievals from LTM are made with cues taken from the focus and the resulting associations are written to other registers in the focus. The contents of still other registers in the focus are stored in (written to) associative memory as we will see below.

Emotions

IDA's design includes mechanisms for emotions (McCauley & Franklin, 1998). She may 'experience' such emotions as guilt at not getting a sailor's orders out on time, frustration at not understanding a message, and anxiety at not being able to convince a sailor to accept a suitable job. Action selection will be influenced by emotions via their effect on drives (see below), modelling work on human action selection (Damasio, 1994). Emotions also influence attention, the means of bringing certain content to consciousness. They also influence the strength with which content is written to LTM. The emotion module has not been implemented in IDA, though it has in earlier 'conscious' software agents (McCauley *et al.*, 2000).

Consciousness mechanism

The apparatus for 'consciousness' consists of a coalition manager, a spotlight controller, a broadcast manager, and a collection of attention codelets whose job it is to bring appropriate contents to 'consciousness' (Bogner *et al.*, 2000). Each attention codelet keeps a watchful eye out for some particular occurrence that might call for 'conscious' intervention. In most cases the attention codelet is watching the workspace, which will likely contain both perceptual information and data created internally, the products of 'thoughts'. Upon encountering such a situation, the appropriate attention codelet will form a coalition with the small number of information codelets that carry the information describing the situation. In the example above of our message, these codelets would carry the sailor's name, his or her social security number, and the message type. This

[2] Not to be confused with focus as in focus of attention, an entirely different concept.

association should lead to the collection of this small number of information codelets, together with the attention codelet that collected them, becoming a coalition. Codelets also have activations. The attention codelet increases its activation in order that the coalition, if one is formed, might compete for the spotlight of 'consciousness'. Upon winning the competition, the contents of the coalition is then broadcast to all codelets. Continuing our example, an attention codelet will note the please-find-job message type, gather information codelets carrying name, social security number and message type, be formed into a coalition, and will compete for 'consciousness'. If or when successful, its contents will be broadcast. Broadcast contents are also stored in (written to) associative memory as the contents of 'consciousness' should be. (This is a correction to earlier descriptions of the model.)

Action selection (decision making)

IDA depends on a behaviour net (Maes, 1989; Negatu & Franklin, 2002) for high-level action selection in the service of built-in drives. She has several distinct drives operating in parallel. These drives vary in urgency as time passes and her environment changes. Behaviours are typically mid-level actions, many depending on several behaviour codelets for their execution. A behaviour net is composed of behaviours, corresponding to goal contexts in GW theory, and their various links. A behaviour looks very much like a production rule, having preconditions as well as additions and deletions. It's typically at a higher level of abstraction often requiring the efforts of several codelets to effect its action. A behaviour can be thought of as the collection of its codelets (processors) in accordance with global workspace theory. Each behaviour occupies a node in a digraph. The three types of links, successor, predecessor and conflictor, of the digraph are completely determined by the pre- and post-conditions of its behaviours (Maes, 1989).

As in connectionist models (McClelland *et al.*, 1986), this digraph spreads activation. The activation comes from that stored in the behaviours themselves, from the environment, from drives, and from internal states. The more relevant a behaviour is to the current situation, the more activation it is going to receive from the environment. Each drive awards activation to those behaviours that will satisfy it. Certain internal states of the agent can also send activation to the behaviour net. One example might be activation from a coalition of codelets responding to a 'conscious' broadcast. Activation spreads from behaviour to behaviour along both excitatory and inhibitory links and a behaviour is chosen to execute based on activation. Her behaviour net produces flexible, tunable action selection for IDA. As is widely recognized in humans the hierarchy of goal contexts is fueled at the top by drives, that is, by primitive motivators, and at the bottom by input from the environment, both external and internal.

Returning to our example, the broadcast is received by appropriate behaviour codelets who know to instantiate a behaviour stream in the behaviour net for reading the sailor's personnel record. They also bind appropriate variables with name and social security number, and send activation to a behaviour that knows how to initiate the accessing of the database. If or when that behaviour is chosen to be executed, behaviour codelets associated with it begin to read data from the sailor's file. This data is written to the workspace. Each such write results in another round of associations, the triggering of an attention codelets, the resulting information coming to 'consciousness', additional binding of variables and passing of activation and the execution of the next behaviour. As long as it's the most important activity going, this process is continued until all the relevant personnel data is written to the workspace. In a similar fashion, repeated runs through 'consciousness' and the behaviour net result in a coarse selection of possible suitable jobs being made from the job requisition database.

The process just described leads us to speculate that in humans, like in IDA, processors (neuronal groups) bring perceptions and thoughts to consciousness. Other processors, aware of the contents of consciousness, instantiate an appropriate goal context hierarchy, which in turn, motivates yet other processors to perform internal or external actions.

Constraint satisfaction

IDA is provided with a constraint satisfaction module designed around a linear functional. It provides a numerical measure of the suitability, or fitness, of a specific job for a given sailor (Kelemen *et al.*, 2002). For each preference (say for San Diego) or issue (say moving costs) or policy (say sea duty following shore duty) there's a function that measures suitability in that respect. Coefficients indicate the relative importance assigned to each preference, issue or policy. The weighted sum measures the job's fitness for this sailor at this time. The same process, beginning with an attention codelet and ending with behaviour codelets, brings each function value to 'consciousness' and writes the next into the workspace. At last, the job's fitness value is written to the workspace.

Deliberation

Since IDA's domain is fairly complex, she requires *deliberation* in the sense of creating possible scenarios, partial plans of actions, and choosing between them (Sloman, 1999). In our example, IDA now has a list of a number of possible jobs in her workspace, together with their fitness values. She must construct a temporal scenario for at least a few of these possible billets to see if the timing will work out (say if the sailor can be aboard ship before the departure date). In each scenario the sailor leaves his or her current post during an approved time interval, spends a specified length of time on leave, possibly reports to a training facility on a certain date, uses travel time, and arrives at the new billet within a given time frame. Such scenarios are valued on how well they fit the temporal constraints (the gap) and on moving and training costs. These scenarios are composed of scenes organized around events, and are constructed in the workspace by the same attention codelet to 'consciousness' to behaviour net to behaviour codelets as described previously.

Voluntary action

We humans most often select actions subconsciously, that is, without conscious thought. But we also make voluntary choices of action, often as a result of the kind of deliberation described above. Baars argues that such voluntary choice is the same a conscious choice (1997, p. 131) We must carefully distinguish between being conscious of the results of an action and consciously deciding to take that action, that is, of consciously deliberating on the decision. It's the latter case that constitutes voluntary action. William James proposed the *ideomotor theory* of voluntary action (James, 1890). James suggests that any idea (internal proposal) for an action that comes to mind (to consciousness) is acted upon unless it provokes some opposing idea or some counter proposal. GW theory adopts James' ideomotor theory as is (Baars, 1988, Chapter 7) and provides a functional architecture for it. The IDA model furnishes an underlying mechanism that implements that theory of volition and its architecture in a software agent.

Suppose that in our example at least one scenario has been successfully constructed in the workspace. The players in this decision making process include several proposing attention codelets and a timekeeper codelet. A proposing attention codelet's task is to propose that a certain job be offered to the sailor. Choosing a job to propose on the basis of the codelet's particular pattern of preferences, it brings information about itself and the proposed job to 'consciousness' so that the timekeeper codelet and others can know of it. Its preference pattern may include several different issues (say priority, moving cost, gap, etc.) with differing weights assigned to each. For example, our proposing attention codelet may place great weight on low moving cost, some weight on fitness value, and little weight on the others. This codelet may propose the second job on the scenario list because of its low cost and high fitness, in spite of low priority and a sizable gap. If no other proposing attention codelet objects (by bringing itself to 'consciousness' with an objecting message) and no other such codelet proposes a different job within a given span of time, the timekeeper codelet will mark the proposed job as being one to be offered. If an objection or a new proposal is made in a timely fashion, it will not do so.

Two proposing attention codelets may alternatively propose the same two jobs several times. Several mechanisms tend to prevent continuing oscillation. Each time a codelet proposes the same job it does so with less activation and so has less chance of coming to 'consciousness'. Also, the timekeeper loses patience as the process continues, thereby diminishing the time span required for a decision. Finally, the metacognitive module watches the whole process and intervenes if things get too bad (Zhang *et al.*, 1998a). A job proposal may also alternate with an objection, rather than with another proposal, with the same kinds of consequences. These occurrences may also be interspersed with the creation of new scenarios. If a job is proposed but objected to, and no other is proposed, the scenario building may be expected to continue yielding the possibility of finding a job that can be agreed upon.

Negotiation

After IDA has selected one or more jobs to be offered to a given sailor, her next chore is to negotiate with the sailor until one job is decided upon. The US Navy is quite concerned about retention of sailors in the service. This depends heavily on the sailor's job satisfaction. Thus the Navy gives a high priority to the assignment of a job that both satisfies the sailor's preferences and offers opportunity for advancement, sometimes including additional training. Whenever possible the final job assignment is made with the sailor's agreement. IDA must negotiate this agreement with the sailor.

When the initial job offerings are made the sailor may respond in several different ways. He may accept one of the jobs offered. He may decline all of them and request some different job assignment. He may ask for a particular job not among those offered. He may ask that the process be postponed until a new requisition list appears, hoping to find something more to his liking. IDA may accede to or deny any of these requests, the decision often dependent on time constraints and/or the needs of the service. The continuing negotiations offer many possible paths. It ends with one job being assigned to the sailor, most often with his agreement, but sometimes without.

IDA must be able to carry out such negotiations. This requires making decisions and responding to the sailor's messages. We've already had a close look at her decision making processes involving both 'consciousness', the behaviour net and voluntary action.

Metacognition

Metacognition guides people to select, evaluate, revise, and abandon cognitive tasks, goals, and strategies (Hacker, 1997). Most researchers agree that metacognition should include knowledge of one's own knowledge and cognitive processes, and the ability to actively monitor and consciously regulate them. While metacognition isn't implemented in the current running version of IDA, it's very much a part of the conceptual IDA model, and has been implemented previously in an earlier 'conscious' software agent, CMattie (Zhang *et al.*, 1998a). This admittedly impoverished metacognition module was able to interrupt oscillations during voluntary action selection, and also to tune the behaviour net by adjusting it's parameters so that it ran more smoothly. Metacognition is not yet conscious in our model. We've experimented with several mechanisms for metacognition including a classifier system, a fuzzy logic controller and a fuzzy classifier system. All worked about equally well. The latter also learned.

Learning

The IDA model incorporates several different learning mechanisms. The simplest is the associative learning that occurs as the contents of 'consciousness' is stored in associative memory with every 'conscious' broadcast. It's the associations between the various items comprising the 'conscious' contents that's learned. Also, associations with other similar items are learned by means of the sparse distributed memory mechanism.

Learning into associative memory can be considered a type of declarative learning. IDA is also capable of a more procedural learning. Codelets in IDA participate in a pandemonium theory style of organization (Jackson, 1987). Those codelets who share time in the spotlight of 'consciousness' have associations between them formed or strengthened, or perhaps diminished if things are going badly. Those codelets sharing time in the playing field also change associations, but at a much lesser rate. This dynamic association of codelets allows a Hebbian type of learning. As codelets become more or less associated, the likelihood of their coming to 'consciousness' together changes, and with it the likelihood of their together helping to initiate some behaviour. These changes in likelihood constitute a kind of procedural learning.

When the same coalition of codelets, acting in parallel or in sequence, often produce a useful result, this coalition can be expected to merge into a higher-level concept codelet. This merging constitutes a second form of temporal proximity learning. The concept codelet, when active, performs the same actions as do its member codelets combined, but without requiring successive executions of the behaviour net. This process is comparable to chunking in SOAR (Laird *et al.*, 1987).

In the conceptual model of IDA we include mechanisms for emotions (McCauley & Franklin, 1998). Action selection will be influenced by emotions via their effect on drives, modelling recent work on human action selection (Damasio, 1994). IDA's emotional mechanism includes an activation passing network with weighted links that allows learning. Weights will decay with disuse. Each use tends to decrease the decay rate. Weights will increase according to a sigmoidal function of any activation carried over the link, allowing for Hebbian style learning. Thus IDA is, in principle, able to learn to use emotions to affect various cognitive functions as we humans do.

Also, IDA is, in principle, able to learn metacognitively using the built-in fuzzy classifier mechanism. In particular, IDA should be able to learn to tune her action selection by means of adjusting the parameters of her behaviour net as CMattie did before her.

Plans are afoot for IDA to learn by being told by a human (Ramamurthy *et al.*, 1998a,b; 2001), but we don't claim this form of learning as part of the IDA conceptual model since the required mechanisms are not yet completely designed.

A final form of learning in IDA allows the automatization of repeated actions as we humans do. Here, the mechanisms are in place and constitute part of the IDA conceptual model. This is new work that will be the subject of another report, and cannot be briefly described here.

'Conscious' and Unconscious Decision Making in IDA

As humans do, IDA uses 'consciousness' during her decision making (action selection) process in several different ways. Also, she makes some decisions unconsciously. In this section we'll describe these various processes.

If I feel hungry and choose to make myself a Nathan's hotdog, I've engaged in *internally generated voluntary action selection* as described by William James' ideomotor theory. Of course, the process may have been much more complex including consciously considering the possibilities of going to Bozo's for barbecue or to Mikasa for sushi instead, or of simply settling for a piece of fruit, or of eating later after finishing the section I'm writing. Once that decision is made a sequence of actions typically follow. Some of these may involve other instances of internally generated voluntary actions, such as the decision of whether to cook the hotdog on the stove or in the microwave, or whether to wear a coat. Such decisions are characterized by a *conscious consideration of alternatives motivated by some internal drive or goal*. The conscious consideration may be of a single alternative (Nathan's hotdog) where the choice is whether to do it.

James' theory, and our implementation of it in IDA, is described in the section on voluntary action above. IDA engages in internally generated voluntary action selection when deciding which jobs to offer the sailor. This action is motivated by her drive to find a job suitable to both the sailor and the US Navy. Since alternatives are considered without being acted upon, the process is a deliberative one in the sense of Sloman (1999).

Voluntary action selection can also be occasioned by external environmental situations. I want to get into a trailer. The door is locked. Shall I try to find the hidden key or shall I look for Shelby who has a key? This is a example of *externally generated voluntary action selection* which is characterized by voluntary action selection motivated by a drive or goal, but occasioned by an obstacle or an opportunity presented by the external environment. IDA's current incarnation doesn't display externally generated voluntary action selection, but she easily could. Suppose IDA was unable to find a sailor's personnel record. She could consider writing to the sailor to check the social security number, the index of the personnel file. Or, she might think of trying a query based on the sailor's name, or she might refer the matter to a human. To implement this capability would require the addition of a few attention codelets and a couple of new behaviour streams with their behaviour codelets. Ideally, the IDA conceptual model should allow for internal construction of these codelets and streams as we humans do. This illustrates one of the many gaps in the model. We're working on this one.

Many of our actions, though not voluntary, are nonetheless consciously mediated. Having retrieved the key for the trailer I must be conscious of at least the steps leading up to the door. I must be conscious of the position of the keyhole. These should not, however, be construed as conscious decision points requiring voluntary action selection. I do not consider whether or not to take the first step, nor whether to insert the key into the keyhole. These actions are parts of several preplanned and executed action sequences, *consciously mediated* in order to fill in required data from the environment, in this case location of step and keyhole. Consciously mediated action selection is involuntary in that a sequence of actions is selected, perhaps voluntarily, and then proceeds via a sequence of action decisions that may involve consciousness but that involvement is not a matter of deciding whether to take the action in question. Rather consciousness provides information needed as a precondition to taking the action, such as a location. Consciousness mediates in helping to align the key with the keyhole.

Consciousness can also mediate internal action selections. Here's an example thoughtfully provided by my daughter Sunny. She and her friend, Christina, were playing the game of guessing people's star signs. Christina made a guess for Sunny's brother, Bruce. As part of the process of seeing if Christina guessed right, Sunny had to recall Bruce's birthday, an internal action, but not voluntarily selected. Sunny wasn't considering whether to remember Bruce's birthday. She simply had to have the birthday consciously available in order to tell it to Christina.

The process of 'consciously' mediated action selection in IDA is implemented as described in the section on action selection above. The preplanned action sequences mentioned above are implemented as behaviour streams. Behaviour codelets instantiate such behaviour streams, perhaps motivated by voluntary action selection, perhaps not. The selection of a behaviour in a stream for execution is accomplished by the behaviour net quite unconsciously, as described above. For example, suppose the executed behaviour reads a field from the sailor's personnel record. The contents of that field, brought to 'consciousness' by an attention codelet, will serve to trigger the reading of the next field, provided the contents aren't unexpected. Behaviour codelets responsible for obtaining the current duty station, seeing the contents of the name field in 'consciousness', know that it's their turn to act. They send (environmental) activation to the next behaviour in the stream, their behaviour, helping it to be executed next. In this case the 'conscious' contents serve as a cue for the next action. We hypothesize that well rehearsed action sequences in humans work the same way.

IDA also performs actions 'consciously' mediated internally. For instance, when composing an email message to a sailor, IDA 'consciously' fills in the blanks in scripts carried by codelets. The intended content of a field is 'consciously' retrieved from working memory by an executing behaviour. Behaviour codelets responding to the resulting 'conscious' broadcast not only send activation to the next behaviour in the stream (fill in this particular field) but bind the variable containing the intended content of the field. Just as one can't insert a key into a key into a keyhole without seeing to align it properly, one can't fill in a blank in a script without knowing the intended content of the blank.

Finally, both humans and IDA, at least in her conceptual model, perform actions that are not consciously mediated. Once the key is aligned with the keyhole, the sequence of actions involving inserting the key and turning it can be done unconsciously, as long as it proceeds as expected. Such action sequences involving primitive actions by a number of distinct muscle groups are typically first learned consciously and become automatized with use. The same process is available to IDA in her conceptual model. Transferring needed data from an external personnel file to her workspace (working memory) is at first done 'consciously' mediated for each piece of data, as described above. Later association between adjacent behaviours become strong enough for one to call the next behaviour in pandemonium style (Jackson, 1987) without the benefit of 'conscious' intervention. Thus we have a machine version of the automatization process. This version also includes the capability of bringing a behaviour back to 'consciousness' when it doesn't work as expected. This work is as yet unpublished.

Is IDA Conscious?

Clearly IDA is quite a complex software agent that models a broad swath of human cognition including 'consciousness' in the sense of implementing global workspace theory (Franklin & Graesser, 1999). But, is there any sense in which she can be said to be conscious without the quotes. Well, there's the ill-defined cluster concept of functional consciousness. IDA exhibits both external and internal voluntary actions selection, as well as consciously mediated action selection of both the internal and external variety. Though IDA does not, as yet, engage in non-routine problem solving, work on adding that capability is in progress. She uses her 'consciousness' module to handle routine problems with novel content. All this together makes a strong case, in my view, for functional consciousness.

But what about phenomenal consciousness? Can we claim it for IDA? Is she *really* a conscious artifact? I can see no convincing arguments for such a claim. One might argue that anything that senses must have some sort of subjective consciousness as a consequence. Would we thus attribute consciousness to a thermostat? Not lightly. On the other hand, I can see no convincing arguments against a claim for phenomenal consciousness in IDA. Yes, she's not biological, much less the possessor of neuropil. Does that really matter? Perhaps. Or, perhaps neuropil is only necessary for phenomenal consciousness in biological agents. IDA is a piece of running software and is thus immaterial. What is there to be conscious? (Tom Ziempke, personal communication) On the other hand, perhaps human consciousness runs as much on a virtual machine as IDA does. These requirements that conscious beings be material or biological or have neuropil may be no more valid than those prompting the denial of phenomenal consciousness to other apes on the grounds that they are not human.

But, can IDA report her 'conscious experiences'? No, but it's only because we haven't built in that capability for her. To have her report the contents of her 'consciousness' would require no new ideas or mechanisms. It would only be a matter of the time and effort to do the work. We know how, and it would not be in principle different from what IDA does now. Reportability is *not* the issue. It's currently part of the IDA conceptual model.

What about self-consciousness? Though a self is postulated as part of global workspace theory, it has not been implemented in IDA, partially because we just didn't understand the concept well enough to know how to implement it. Damasio's formulation of a self (Damasio, 1999) has given us ideas so that adding a self to IDA by way of a proto-self now seems possible. However the mechanisms are not in place, meaning that a self is not yet part of the IDA conceptual model. It does seem at least possible that a self is necessary for phenomenal consciousness though self-consciousness isn't. The work of Humphrey, speculating on how phenomenal consciousness might have evolved in relatively simple organisms, makes this idea seem quite plausible (Humphrey, 2000).

Conclusion

So, what are we to conclude from all this? Suppose the neuroscientists succeed in finding neural correlates of phenomenal consciousness. This may yield evidence as to the consciousness or lack thereof of various animals. It seems unlikely to help us at all with the question of machine consciousness. Nor does anything else seem likely to be of help. Subjective consciousness seems to be such an inherently first-person phenomenon as to be impervious to any form of proof. A solipsistic position seems as defensible as any other. The best we can do is to offer evidence in favour such as reportablity that can be externally confirmed. Such evidence, whether based on behaviour or on internal structure and/or activity, is unlikely to make much of an impression on the philosophers who are willing to take seriously the possibility of zombies, thus giving rise to the hard problem of consciousness.

Following Chalmers (1996) I'm beginning to view phenomenal consciousness as a fundamental process of nature comparable to mass or energy. Both mass and energy can only be measured indirectly by their effects though inferences can be drawn from their structure and activity (again effects). I suspect that the same will prove true of phenomenal consciousness. How can we measure its effects? Will this become the *important problem* of consciousness? Likely, I would think. In the realm of machine consciousness this problem seems even harder. Deducing structure from behaviour is notoriously difficult, to which both the psychologists and the neuroscientists will attest. In software agents such deductions are both theoretically impossible and practically problematic. Still, measuring mass and energy weren't easy problems until someone solved them.

Where does that leave us? At least for now, the attribution of phenomenal consciousness will depend on the level of evidential support the person doing the attributing will accept. Were I faced with Commander Data from *Star Trek: The Next Generation* I would undoubted attribute consciousness. On the other had I do not attribute phenomenal consciousness to my own 'conscious' software agent, IDA, in spite of her many human-like behaviours. This in spite of watching several US Navy detailers repeatedly nodding their heads saying 'Yes, that's how I do it' while watching IDA's internal and external actions as she performs her task. I suspect they might be more willing to attribute phenomenal consciousness than I.

Speaking of the 'additional magic ingredient' that produces phenomenal consciousness out of functional consciousness, Harvey says '. . . the additional magic ingredient for [phenomenal consciousness] is merely a change in attitude in us, the observers' (Harvey, 2002). The claim is that the presence or absence of phenomenal consciousness can never be more than a matter of attribution. Bearing that claim in mind, let me close with a prediction. I expect there to be software agents and robots sufficiently intelligent, sufficiently capable, and sufficiently communicative that people with simply assume that they are conscious artifacts (Moravec, 1988). The issue of machine consciousness will no longer be relevant.

Acknowledgements

This work has been supported in part by ONR grant N00014-98-1-0332. Much of it emerged from the combined effort of members of the Conscious Software Research Group at the University of Memphis. These include, or have included, Art Graesser, Lee McCauley, Hongjun Song, Zhaohua Zhang, Satish Ambati, Ashraf Anwar, Myles Bogner, Arpad Kelemen, Ravikumar Kondadadi, Irina Makkaveeva, Aregahegn Negatu, Alexei Stoliartchouk, Uma Ramamurthy.

References

Aleksander, I. (2000), *How to Build a Mind* (London: Weidenfeld and Nicolson).
Allen, J.J. (1995), *Natural Language Understanding* (Redwood City, CA: Benjamin/Cummings).
Anwar, A. and Franklin, S. (to appear), 'Sparse distributed memory for "conscious" software agents', *Cognitive Systems Research*.
Anwar, A., Dasgupta, D. and Franklin, S. (1999), 'Using genetic algorithms for sparse distributed memory initialization', paper presented at *International Conference Genetic and Evolutionary Computation* (GECCO), July, 1999.
Baars, B.J. (1988), *A Cognitive Theory of Consciousness* (Cambridge: Cambridge University Press).
Baars, B.J. (1997), *In the Theater of Consciousness* (New York: Oxford University Press).
Baars, B.J. (2001), 'Surrounded by consciousness: The scientific evidence for animal consciousness since the Mesozoic', paper presented at Consciousness and its Place in Nature: Toward a Science of Consciousness; Skovde, Sweden; August 7–11, 2001.
Baars, B.J. (2002), 'The conscious access hypothesis: Origins and recent evidence', *Trends in Cognitive Science*, **6**, pp. 47–52.
Bogner, M., Ramamurthy, U. and Franklin, S. (2000), '"Consciousness" and conceptual learning in a socially situated agent', in *Human Cognition and Social Agent Technology*, ed. K. Dautenhahn (Amsterdam: John Benjamins).
Chalmers, D.J. (1996), *The Conscious Mind* (New York: Oxford University Press).
Cotterill, R. (2001), 'Consciousness in theory — and in the computer?', paper presented at Consciousness and its Place in Nature: Toward a Science of Consciousness. Skovde, Sweden; August 7–11, 2001.
Damasio, A.R. (1994), *Descartes' Error* (New York: Gosset; Putnam Press).
Damasio, A.R. (1999), *The Feeling of What Happens* (New York: Harcourt Brace).
Edelman, G.M. (1987), *Neural Darwinism* (New York: Basic Books).
Edelman, G.M. and Tononi, G. (2000). *A Universe of Consciousness* (New York: Basic Books).
Ericsson, K. A. and Kintsch, W. (1995), 'Long-term working memory', *Psychological Review*, **102**, pp. 21–245.
Franklin, S. (1995), *Artificial Minds* (Cambridge, MA: MIT Press).
Franklin, S. (1997), 'Autonomous agents as embodied AI', *Cybernetics and Systems*, **28**, pp. 499–520.
Franklin, S. and Graesser, A.C. (1997), 'Is it an agent, or just a program?: A taxonomy for autonomous agents', In *Intelligent Agents III* (Berlin: Springer Verlag).
Franklin, S. and Graesser, A. (1999), 'A software agent model of consciousness', *Consciousness and Cognition*, **8**, pp. 285–305.
Franklin, S., Kelemen, A. and McCauley, L. (1998), 'IDA: A cognitive agent architecture', In *IEEE Conf. on Systems, Man and Cybernetics* (IEEE Press).
Freeman, W.J. (1995), *Societies of Brains* (Hillsdale, NJ: Lawrence Erlbaum).
Freeman, W.J. (1999), *How Brains Make Up Their Minds* (London: Weidenfeld & Nicolson).
Gallup, G. (1982), 'Self-awareness and the emergence of mind in primates', *American Journal of Primatology*, **2**, pp. 237–46.
Glenberg, A.M. (1997), 'What memory is for', *Behavioral and Brain Sciences*, **20**, pp. 1–19.
Hacker, D. (1997), 'Metacognitive: Definitions and empirical foundations', in *Metacognition in Educational Theory and Practice*, ed. D. Hacker, J. Dunlosky and A. Graesser (Hillsdale, NJ: Erlbaum).

Harvey, I. (2002), *Evolving Robot Consciousness: The Easy Problems and the Rest. In Evolving Consciousness*, ed. J. H. Fetzer (Amsterdam: John Benjamins).
Hofstadter, D.R. and Mitchell, M. (1994), 'The Copycat Project: A model of mental fluidity and analogy-making', in *Advances in Connectionist and Neural Computation Theory, Vol. 2: Logical Connections*, ed. K.J. Holyoak and J.A. Barnden (Norwood, NJ: Ablex).
Humphrey, N. (2000), *How to Solve the Mind-Body Problem* (Exeter: Imprint Academic).
Jackson, J.V. (1987), 'Idea for a mind', *Siggart Newsletter*, **181**, pp. 23–6.
James, W. (1890), *The Principles of Psychology* (Cambridge, MA: Harvard University Press).
Jurafsky, D. and Martin, J.H. (2000), *Speech and Language Processing* (Englewood Cliffs, NJ: Prentice-Hall).
Kanerva, P. (1988), *Sparse Distributed Memory* (Cambridge MA: The MIT Press).
Kelemen, A., Liang, Y., Kozma, R. and Franklin, S. (2002), 'Optimizing intelligent agent's constraint satisfaction with neural networks', in *Innovations in Intelligent Systems*, ed. A. Abraham and B. Nath (Heidelberg: Springer-Verlag), pp. 255–72.
Laird, E. J., Newell, A. and Rosenbloom P.S. (1987), 'SOAR: An architecture for general intelligence', *Artificial Intelligence*, **33**, pp. 1–64.
Maes, P. (1989), 'How to do the right thing', *Connection Science*, **1**, pp. 291–323.
Maturana, H.R. (1975), 'The organization of the living: A theory of the living organization', *International Journal of Man-Machine Studies*, **7**, pp. 313–32.
Maturana, R.H. and Varela, F.J. (1980), *Autopoiesis and Cognition: The Realization of the Living* (Dordrecht: Reidel).
McCauley, L. (2002), 'Neural schemas: Toward a comprehensive mechanism of mind', PhD Dissertation. The University of Memphis.
McCauley, T.L. and Franklin, S. (1998), 'An architecture for emotion', in AAAI Fall Symposium *Emotional and Intelligent: The Tangled Knot of Cognition* (Menlo Park, CA: AAAI Press).
McCauley, L., Franklin, S. and Bogner, M. (2000), 'An emotion-based 'conscious' software agent architecture', in *Affective Interactions, Lecture Notes on Artificial Intelligence*, vol. 1814, ed. A. Paiva (Berlin: Springer).
McClelland, J.L., Rumelhart, D.E., *et al.* (1986), *Parallel Distributed Processing, Vol. 1* (Cambridge, MA: MIT Press).
Minsky, M. (1985), *The Society of Mind* (New York: Simon and Schuster).
Moravec, H. (1988), *Mind Children* (Cambridge MA: Harvard University Press).
Negatu, A. and Franklin, S. (2002). 'An action selection mechanism for "conscious" software agents', *Cognitive Science Quarterly*, **2**, pp. 363–86.
Ornstein, R. (1986), *Multimind* (Boston: Houghton Mifflin).
Ramamurthy, U., Bogner, M. and Franklin, S. (1998a), '"Conscious" learning in an adaptive software agent', in *Proceedings of the Second Asia Pacific Conference on Simulated Evolution and Learning*, ed. X. Yao, R.I. McKay, C.S. Newton, J.H. Kim and T. Furuhashi (Canberra: University of New South Wales).
Ramamurthy, U., Franklin, S. and Negatu, A. (1998b), 'Learning concepts in software agents', in *From Animals to Animats 5: Proceedings of The Fifth International Conference on Simulation of Adaptive Behavior*, ed. R. Pfeifer, B. Blumberg, J-A. Meyer and S.W. Wilson (Cambridge, MA: MIT Press).
Ramamurthy, U., Negatu, A. and Franklin, S. (2001), 'Learning mechanisms for intelligent systems', SSGRR-2001 International Conference on Advances in Infrastructure for e-Business, e-Education and e-Science on the Internet. L'Aquila, Italy; August 6–12, 2001.
Sloman, A. (1987), 'Motives, mechanisms, emotions', *Cognition and Emotion*, 1, pp. 217–34.
Sloman, A. (1995), 'A philosophical encounter', in *Proceedings of the Fourteenth International Joint Conference on Artificial Intelligence* (Morgan Kaufman).
Sloman, A. (1999), 'What sort of architecture is required for a human-like agent?' in *Foundations of Rational Agency*, ed. M. Wooldridge and A. Rao (Dordrecht: Kluwer Academic Publishers).
Zhang, Z., Dasgupta, D. and Franklin, S. (1998a), 'Metacognition in software agents using classifier systems', in *Proceedings of the Fifteenth National Conference on Artificial Intelligence* (Madison, WI: MIT Press).
Zhang, Z., Franklin, S., Olde, B., Wan, Y. and Graesser, A. (1998b), 'Natural language sensing for autonomous agents', in *Proceedings of IEEE International Joint Symposia on Intellgence Systems 98*.

Stevan Harnad

Can a Machine Be Conscious? How?

The Nature of the Question

Asking whether a machine can be conscious is rather like asking whether one has stopped beating one's wife: The question is so heavy with assumptions that either answer would be incriminating!

The answer, of course, is: It depends entirely on what you mean by 'machine'! If you mean the current generation of man-made devices (toasters, ovens, cars, computers, today's robots), the answer is: almost certainly not.

Empirical risk

Why 'almost'? Two reasons, the first being the usual one: (1) 'empirical risk.' We know since at least Descartes that even scientific 'laws' are merely very probable, not certain. Only mathematical laws — which describe consequences that follow provably (i.e., on pain of contradiction) from our own assumptions — are *necessarily* true. But this certainty and necessity are unnecessary for physics; almost-certainty will do. 'Can a particle travel faster than the speed of light?' Almost certainly not (at least on current-best theory — or at least the last word of it that trickled down to this non-physicist). '*Could* a particle travel faster than light?' That certainly is not provably impossible, though it might be impossible given certain assumptions. (But those assumptions are not necessarily correct.)

The other-minds problem

Empirical risk besets all scientific hypotheses, but let us agree that it is not something we will worry about here. There is no need for roboticists to be holier than physicists. The second reason for the 'almost' is peculiar to robotics, however, and it is called (2) the 'other minds problem' (Harnad, 1991): There is no way to be certain that any other entity than myself is conscious (I am speaking deictically: please substitute yourself for me, if you too are conscious). This, too, we owe to Descartes.

In the long history of philosophy the other-minds problem has been puzzled over for a variety of reasons, usually variants on questions about what one can and cannot know for sure. These epistemic questions are interesting, but we will not worry about them here, for the usual reason, which is the following: It looks on the face of it as if the right strategy for handling the other-minds problem is identical to the strategy for handling empirical risk, namely, to note that although we can only be 100% certain about two things — about (1) mathematics and about (2) our own consciousness — all else being just a matter of probability, some things, such as scientific laws and the consciousness of our fellow human beings, are nevertheless so close to 100% sure that it is a waste of time worrying about them. (Let us also note, though, that in the empirical science of robotics, that extra layer of risk that comes from the other-minds problem might just come back to haunt us.)

So let us agree not to worry about the other-minds problem for now: People other than myself are almost certainly conscious too, and toasters and all other human artefacts to date are almost certainly not.

What Is a Machine?

Have we now answered the question 'Can a machine be conscious?' It sounds as if, at the very least, we have answered the question 'Is any machine we have built to date conscious?' That makes the original question sound as if it was only asking about what we can and cannot build, which is like asking whether we can build a rocket that can reach Alpha Centauri (a rather vague and arbitrary question about quantitative limitations on future technology). But is 'machine' *defined* as 'what human beings can build'? I think that defines 'artefact' — but 'machine'? Or, rather, do we really want to ask merely: 'Can a man-made artefact be conscious?' And even that would be rather vague, for 'man-made' is itself rather vague. Common sense dictates that human procreation does not count as 'man-making' in this context. But what about genetic or other biological engineering? If the day comes when we can craft organisms, even humans, molecule by molecule, in the laboratory, does anyone — or rather, anyone who has agreed to discount the other-minds problem when it comes to naturally crafted fellow-humans — doubt that such a bottom-up construction of a clone would be conscious too?

So 'man-made' is a wishy-washy term. It does not pick out what we mean by 'machine' here. Surely a toaster (that very same device) would not become *more* eligible for consciousness if it happened to grow on a tree instead of being fabricated by one of us. By the same token, a toaster would not become any *less* of a 'machine' (whatever that turns out to mean) by growing on a tree: Two toasters, identical right down to the last component, one of which I built and the other of which grew on a tree, are surely both 'machines' (whatever that means) if either one of them is. Another way to put this is that we need a definition of 'machine' that is strictly structural/functional, and not simply dependent on its historic origins, if we want to make our question about what machines can and cannot do (or be) into a substantive rather than an arbitrary one.

Kinds of machines

But I am afraid that if we do follow this much more sensible route to the definition of 'machine' we will find that a machine turns out to be simply: *any causal physical system*, any 'mechanism'. And in that case, biological organisms are machines too, and the answer to our question 'Can a machine be conscious' is a trivial 'Yes, of course'. *We* are conscious machines. Hence, machines can obviously be conscious. The rest is just about what *kinds* of machines can and cannot be conscious, and *how* — and that becomes a standard empirical research programme in 'cognitive science': The engineering side of cognitive science would be the forward engineering of man-made conscious systems and the biological side of cognitive science would be the reverse engineering of natural conscious systems (like ourselves, and our fellow organisms): figuring out how our brains work.

Except for one problem, and it is the one that risked coming back to haunt us: What does it mean to 'forward engineer' (or, for that matter, to 'reverse engineer') a conscious system? It is to give a causal explanation of it, to describe fully the inner workings of the mechanism that gives rise to the consciousness.

Forward- and reverse-engineering the heart and brain

Let us take a less problematic example: To forward engineer a cardiac system (a heart) is to build a mechanism that can *do* what the heart can *do*. To reverse engineer the heart is to do the same thing, but in such a way as to explain the structure and the function of the biological heart itself, and not merely create a prosthesis that can take over some its function. Either way, the explanation is a structural/ functional one. That is, both forward and reverse engineering explain everything that a heart can *do*, and how, whereas reverse engineering goes on to explain what the heart *is* (made out of) and how it in particular happens to do what hearts can do.

Now let us try to carry this over to the brain, which is presumably the organ of consciousness. To forward engineer the brain is to build a mechanism that can do what the brain can do; to reverse engineer it is to do the same thing, but in such a way as to explain the structure and function of the biological brain itself. Either way, the explanation is a structural/functional explanation; that is, it explains what the brain is (made out of), and what it can do: how it works.

How does the ghost of the other-minds problem spoil this seemingly straightforward extension of the cardiac into the cogitative? First, consider the forward engineering: If we were forward engineering cardiac function, trying to build a prosthesis that took over doing all the things the heart does, we would do it by continuing to add and refine functions until we built something that was functionally indistinguishable from a heart. (One test of our success might be whether such a prosthesis could be implanted into humans from cradle to grave with no symptom that it was missing any vital cardiac function.) This forward-engineered cardiac system would still be structurally distinguishable from a natural heart, because it had omitted other properties of the heart — non-cardiac ones, but biological properties nonetheless — and to capture those too may require reverse engineering of the constructive, molecular kind we mentioned earlier, building it bottom-up out of biological components.

The thing to note is that this cardiac research program is completely unproblematic. If a vitalist had asked 'Can a machine be cardiac?' we could have given him the sermon about 'machines' that we began with (i.e., you should instead be asking 'What *kind* of machine can and cannot be cardiac, and how?'). Next, we could have led him to the complete reverse-engineered heart, our constructed cardiac clone, using mechanistic principles (i.e., structure, function and causality) alone. At no point would the cardiac vitalist have any basis for saying: 'But how do we know that this machine is *really* cardiac?' There is no way left (other than ordinary empirical risk) for any difference even to be *defined*, because every structural and functional difference has been eliminated in the complete reverse-engineered heart.

The same would be true if it had been life itself and not just cardiac function that had been at issue: If our question had been 'Can a machine be alive?', the very same line of reasoning would show that there is absolutely no reason to doubt it (apart from the usual empirical risk, plus perhaps some intellectual or technological doubts of the Alpha Centauri sort). Again, the critical point is when we ask of the man-made clone: 'But how do we know that this machine is *really* alive?' If there are two structurally and functionally indistinguishable systems, one natural and the other man-made, and their full causal mechanism is known and understood, what does it even *mean* to ask 'But what if one of them is really alive, but the other is not?' What property is at issue that one has and the other lacks, when all properties have already been captured by the engineering (Harnad, 1994)?

Animism at the heart of vitalism

Yet this last worry — 'How can we know it's alive?' — should sound familiar. It sounds like the other-minds problem. Indeed, I suspect that, if we reflect on it, we will realize that it *is* the other-minds problem, and that what we are really worrying about in the case of the man-made system is that there's nobody home in there, there is no ghost in the machine, and that ghost, as usual, is consciousness. That's the property that might be missing.

So, chances are that it was always animism that was at the heart of vitalism. Let us agree to set vitalism aside, however, as there is certainly no way we can know whether something can be alive yet not conscious (or incapable of returning to consciousness). Plants and micro-organisms and irreversibly comatose patients will always be puzzles to us in that respect. So let us not dwell on these inscrutable cases and states. Logic already dictates that any vitalist who does accept that plants are *not* conscious would be in exactly the same untenable position if he went on to express scepticism about whether the complete artificial plant is really alive as the sceptic about the complete artificial heart (worried about whether it's really a heart): If there's a difference, what's the difference? What vital property is at issue? If you can't find one (having renounced on consciousness itself), then you are defending an empty distinction.

Turing-Testing

The same is most definitely *not* true in the case of worries about consciousness itself. Let us take it by steps. First, we forward engineer the brain: We build a robot that can pass the Turing test (Turing, 1950; Harnad, 1992): It can *do* everything a real human can do, for a lifetime, indistinguishably from a real human (except perhaps for appearance: we will return to that).

Let us note, though, that this first step amounts to a tall order, probably taller than the order of getting to Alpha Centauri. But we are talking about 'can' here, that is, about what is possible or impossible (for a machine), and how and why, rather than just what happens to be within our actual human technological reach.

Doing vs. feeling: the feeling/function problem

So, supposing we do succeed in building such a Turing-scale robot (we are no longer talking about toasters here). Now, the question is whether he is really conscious: On the face of it, the only respect in which he is really indistinguishable from us in everything he can *do*. But conscious is something I *am*, not something I *do*. In particular, it is something I *feel*; indeed, it is the *fact* that I feel. So when the sceptic about that robot's consciousness — remember that he cannot be a sceptic about machine consciousness in general: we have already eliminated that by noting that people are a kind of machine, too — wants to say that the robot is the wrong *kind* of machine, that he lacks something essential that we humans have, we all know exactly what difference the sceptic is talking about, and it certainly is not an empty difference. He is saying that the robot does not *feel*, it merely behaves as if it feels — behaves exactly, indeed Turing-indistinguishably, as if it feels, but it does not feel.

Empirical robotics

It is time to remind ourselves of why it is that we agreed to set aside the other-minds problem in the case of our fellow human beings: Why is it that we agreed not to fret over whether other people really have minds (as opposed to merely acting just as if they had minds, but in reality being feelingless Zombies)? It was for the same kind of reason that we don't worry about empirical risk: Yes, it could be that the lawful regularities that nature seems to obey are just temporary or misleading; there is no way to prove that tomorrow will be like today; there is no way to prove that things are as they appear. But there is no way to act on the contrary either (as long as the empirical regularities keep holding).

Empirical risk is only useful and informative where there is still actual uncertainty about the regularities themselves: where it is not yet clear whether nature is behaving as if it is obeying *this* law or *that* law; while we are still trying to build a causal explanation. Once that is accomplished, and all appearances are consistently supporting this law rather than that one, then fretting about the possibility that despite all appearances things might be otherwise is a rather empty exercise. It is fretting about a difference that makes no difference.

Of course, as philosophers are fond of pointing out, our question is (or should be) an *ontic* one, about what really does and does not exist, what really is and is not true, rather than merely an *epistemic* one, about what we can and cannot know, what is and is not 'useful and informative', what does or does not make an empirical difference to us. Epistemic factors (what's knowable or useful to know) have absolutely no power over ontic ones (what there is, what is true).

It would be wise for mere cognitive scientists to concede this point. Just as it is impossible to be certain that the laws in accordance with which nature seems to behave are indeed the true laws of nature, it is impossible to be certain that systems that *behave as if* they feel, truly feel. Having conceded this point regarding *certainty*, however, only a fool argues with the Turing-indistinguishable: Yes, the true laws could be other than the apparent laws, but if I can't tell the two apart empirically, I'd best not try to make too much of that distinction! By the same token, a robot that is indistinguishable for a lifetime from a feeling person might be a zombie, but if I can't tell the two apart empirically, I'd best not try to make too much of that distinction (Harnad, 2000).

Indistinguishable? But surely there are plenty of ways to distinguish a robot from a human being. If you prick us, do we not bleed? So perhaps the sceptic about the forward-engineered robot should hold out for the reverse-bioengineered one, the one made out of the right stuff, Turing- indistinguishable both inside and out, and at both the macro and micro levels. It is only about that machine that we can reply to our reformulated question — 'What kinds of machine can and cannot be conscious?' — that only *that* kind can.

But would we be right, or even empirically or logically justified in concluding that? To put it in a more evocative way, to highlight the paradoxical polarity of the 'risks' involved: Would we be morally justified in concluding that whereas the reverse-bioengineered machines, because they are empirically indistinguishable from natural machines like us, clearly cannot be denied the same human rights as the rest of us, the forward-engineered machines, because they are merely Turing-indistinguishable from us in their (lifelong) behavioural capacity *can* be safely denied those rights and treated as unfeeling Zombies (toasters)?

Other-mind reading and Turing-testing

To answer this question we need to look a little more closely at both our empirical methodology in cognitive science and our moral criteria in real life. Let us consider the second first. Since at least 1978 (Premack, 1978) there has grown an area of research on what is sometimes called 'theory of mind' and sometimes 'mind-reading' in animals and children. This work is not a branch of philosophy or parapsychology as it might sound; it is the study of the capacity of animals and children to detect or infer what others 'have in mind'. (As such, it should really be called research on 'other-mind perception'.)

It has been found that children after a certain age, and certain animals, have considerable skill in detecting or inferring what others (usually members of their own species) are feeling and thinking (Whiten, 1991; Baron-Cohen, 1995). The propensity for developing and exercising this mind-reading skill was probably selected for by evolution, hence is inborn, but it also requires learning and experience to develop. An example of its more innate side might be the capacity to understand facial expressions, gestures and vocalizations that signal emotions or intentions such as anger and aggression; a more learning-dependent example might be the capacity to detect that another individual has seen something, or wants something, or knows something.

Let us note right away that this sort of mind-reading is a form of Turing-testing: inferring mental states from behaviour. The ‘behaviour’ might be both emitted and detected completely unconsciously, as in the case of the release and detection of pheromones, or it might be based on very particular conscious experience, as when I notice that you always purse your lips in a certain way when you think I have lied to you. And there is everything in between; my sense of when you are agitated vs. contented, along with their likely behavioural consequences, might be a representative midpoint. Language (which, let us not forget, is also a behaviour) is probably the most powerful and direct means of mind-reading (Harnad, 1990; Cangelosi & Harnad, 2000).

Hence, apart perhaps from direct chemical communication between brains, all mind-reading is based on behaviour: Turing-testing. It could hardly have been otherwise. We know, again since at least Descartes, that the only mind we can read *other* than by Turing-testing is our own. As far as other minds are concerned, absent genuine telepathic powers (which I take to be a fiction, if not incoherent), the only database available to us for other-mind-reading is other-bodies’ behaviour.

We do have to be careful not to make the ontic/epistemic conflation here: This does not mean that all there is to mind is behaviour (as the blinkered behaviourists thought)! But it does mean that the only way to read others’ minds is through their behaviour, i.e., through Turing-testing.

Functional- vs. structural/functional-indistinguishability

Now, back to our two robots, the reverse-bioengineered one to whom we were ready to grant human rights and the merely forward-engineered one about whom we were not sure: Both are Turing-indistinguishable from us behaviourally, but only the first is anatomically correct. We’re all machines. Is only the first one the right *kind* of machine to have a mind? On what basis could we possibly conclude that? We have ascertained that all mind-reading is just behaviour-based Turing-testing, and all three of us (the two man-made robots and me) are indistinguishable in that respect. What else is there? The rest of the neuromolecular facts about the brain? Which facts?

There are countless facts about the brain that could not possibly be relevant to the fact that it has a mind: its weight, for example. We know this, because there is a huge range of variation in human brain mass — from the massive brain of a huge man to the minute brain of a microcephalic, who nevertheless feels pain when he is pinched. Now imagine trying to narrow down the properties of the brain to those that are necessary and sufficient for its having a mind. This turns out to be just another variant of our original question: ‘What kinds of machines can and cannot be conscious?' We know brains can be, but how and why? What are their relevant properties (if their weight, for example, is not)? Now imagine paring down the properties of the brain, perhaps by experimenting with bioengineered variations, in order to test which ones are and are not conscious. What would the test be?

Turing-filtering relevant brain function

We are right back to Turing-testing again! The only way to sort out the relevant and irrelevant properties of the biological brain, insofar as consciousness is concerned, is by looking at the brain's behaviour. That is the only non-telepathic methodology available to us, because of the other-minds problem. The temptation is to think that 'correlations' will somehow guide us: Use brain scanning to find the areas and activities that covary with conscious states, and those will be the necessary and sufficient conditions of consciousness. But how did we identify those correlates? Because they were correlates of behaviour. To put it another way: When we ask a human being (or a reverse-bioengineered robot) 'do you feel this?' we believe that the accompanying pattern of activity is conscious because we believe him when he says (or acts as if) he feels something — not the other way round: It is not that we conclude that his behaviour is conscious because of the pattern of brain activity; we conclude that the brain activity is conscious because of the behaviour.

So, by the same token, what are we to conclude when the forward-engineered robot says the same thing, and acts the same way (across a lifetime)? If we rely on the Turing criterion in the one case and not the other, what is our basis for that methodological distinction? What do we use in its place, to conclude that this time the internal correlates of the very same behaviour are *not* conscious states?

The answer to our revised question — 'What kinds of machines can be conscious (and how)?' — has now come into methodological focus. The answer is: The kinds that can pass the Turing test, and by whatever means are necessary and sufficient to pass the Turing test.

Darwin and telepathy

If we have any residual worries about zombies passing the Turing test, there are two ways to console ourselves. One is to remind ourselves that not even the Blind Watchmaker who forward engineered us had a better way: survival and reproduction are just Turing functions, too: Darwin is no more capable of telepathy than we are. So there is no more (or less) reason to worry that zombies could slip through the Turing filter of evolution than that they could slip through the Turing filter of robotic engineering (Harnad, 2002).

Turing and telekinesis

Our second consolation is the realization that the problem of explaining how (and why) *we* are not zombies (Harnard, 1995) — otherwise known as the 'mind/body problem' — is a 'hard' problem (Shear, 1997), and not one we are ever likely to solve. It would be easy if telekinetic powers existed: Then feelings would be physical forces like everything else. But there is no evidence at all that feelings are causal forces. That is why our forward- and reverse-engineering can only explain how it is that we can *do* things, not how it is that we can *feel* things. And that is why the ghost in the machine is destined to continue to haunt us even after all cognitive science's empirical work is done (Harnad, 2001).

References

Baron-Cohen, S. (1995), *Mindblindness: An Essay On Autism and Theory Of Mind* (Cambridge, MA: MIT Press).

Cangelosi, A. and Harnad, S. (2001), 'The adaptive advantage of symbolic theft over sensorimotor toil: Grounding language in perceptual categories', *Evolution of Communication*, 4 (1), pp. 117–42. http://cogprints.soton.ac.uk/documents/disk0/00/00/20/36/index.html

Harnad, S. (1990), 'The symbol grounding problem', *Physica D*, **42**, pp. 335–46. http://cogprints.soton.ac.uk/documents/disk0/00/00/06/15/index.html

Harnad, S. (1991), 'Other bodies, other minds: A machine incarnation of an old philosophical problem', *Minds and Machines*, **1**, pp. 43–54. http://cogprints.soton.ac.uk/documents/disk0/00/00/15/78/index.html

Harnad, S. (1992), 'The Turing Test is not a trick: Turing indistinguishability is a scientific criterion', *SIGART Bulletin*, **3** (4) (October 1992), pp. 9–10. http://cogprints.soton.ac.uk/documents/disk0/00/00/15/84/index.html

Harnad, S. (1994), 'Levels of functional equivalence in reverse bioengineering: The Darwinian Turing Test for artificial life', *Artificial Life*, **1** (3), pp. 293–301. http://cogprints.soton.ac.uk/documents/disk0/00/00/15/91/index.html

Harnad, S. (1995), 'Why and how we are not zombies', *Journal of Consciousness Studies*, **1** (2), pp. 164–7. http://cogprints.soton.ac.uk/documents/disk0/00/00/16/01/index.html

Harnad, S. (2000), 'Minds, machines, and Turing: The indistinguishability of indistinguishables', *Journal of Logic, Language, and Information*, **9** (4), pp. 425–45 (special issue on 'Alan Turing and Artificial Intelligence'). http://cogprints.soton.ac.uk/documents/disk0/00/00/16/16/index.html

Harnad, S. (2001), 'No easy way out', *The Sciences*, **41** (2), pp. 36–42. http://cogprints.soton.ac.uk/documents/disk0/00/00/16/24/index.html

Harnad, S. (2002), 'Turing indistinguishability and the Blind Watchmaker', in *Evolving Consciousness*, ed. J. Fetzer (Amsterdam: John Benjamins), pp. 3–18. http://cogprints.soton.ac.uk/documents/disk0/00/00/16/15/index.html

Premack, D. and Woodruff, G. (1978), 'Does the chimpanzee have a theory of mind?', *Behavioral & Brain Sciences*, **4**, pp. 515–26.

Shear, J. (ed. 1997), Explaining Consciousness: The 'Hard Problem' (Cambridge, MA: MIT Press).

Turing, A.M. (1950), 'Computing machinery and intelligence', *Mind*, **49**, pp. 433–60. http://cogprints.soton.ac.uk/documents/disk0/00/00/04/99/index.html

Whiten, A. (ed. 1991), *Natural Theories Of Mind: Evolution, Development, and Simulation of Everyday Mindreading* (Oxford: Blackwell).

Owen Holland and Rod Goodman

Robots With Internal Models

A Route to Machine Consciousness?

New organs of perception come into being as a result of necessity — therefore, O man, increase your necessity so that you may increase your perception.

Jalaluddin Rumi, Persian poet (1207–1273)

Introduction

We are engineers, and our view of consciousness is shaped by an engineering ambition: we would like to build a conscious machine. We begin by acknowledging that we may be a little disadvantaged, in that consciousness studies do not form part of the engineering curriculum, and so we may be starting from a position of considerable ignorance as regards the study of consciousness itself. In practice, however, this may not set us back very far; almost a decade ago, Crick wrote: 'Everyone has a rough idea of what is meant by consciousness. It is better to avoid a precise definition of consciousness because of the dangers of premature definition. Until the problem is understood much better, any attempt at a formal definition is likely to be either misleading or overly restrictive, or both' (Crick, 1994). This seems to be as true now as it was then, although the identification of different aspects of consciousness (P-consciousness, A-consciousness, self consciousness, and monitoring consciousness) by Block (1995) has certainly brought a degree of clarification. On the other hand, there is little doubt that consciousness does seem to be something to do with the operation of a sophisticated control system (the human brain), and we can claim more familiarity with control systems than can most philosophers, so perhaps we can make up some ground there.

The starting point we have chosen is at the conjunction of four fairly uncontroversial observations:

- consciousness is known to arise from the operation of the human brain
- of all brains, the human brain has the highest capacity for intelligence

- the human brain evolved from simpler brains
- the human brain is a control system

Taken together, these suggest a strategy: step by step, develop a series of control systems capable of demonstrating increasingly high intelligence, and monitor their performance at each stage to see when and if they show any signs of consciousness. The immediate appeal is that it does not involve sitting down and 'inventing' machine consciousness, or human-level intelligence, either of which would be extremely problematical; the technical aspect of this approach involves nothing more than the successive development of a series of control systems for producing increasingly intelligent behaviour, and that is a reasonable goal for engineers. Of course, it may also be completely misguided, in that high intelligence may not automatically entail consciousness, but, given the present state of ignorance about so much to do with consciousness, it seems a risk worth taking.

This paper explores some of the issues relevant to the consciousness-via-incremental-intelligence programme, and makes two key design decisions: to embody intelligence in a physical robot, and to concentrate on the exploitation of internal models. It also presents some speculations about how the pursuit of intelligence in this way may lead us to a system with at least some of the characteristics of consciousness.

Choosing an Architectural Stance: Internal Modelling

The scheme as outlined so far is too broad to define a distinctive and manageable approach. There are a great many types both of control systems, and of information processing architectures; we cannot explore them all, but the choice of one particular type should be principled rather than arbitrary. It should of course hold out the prospect of offering a route to high intelligence, but ideally it should also combine some credible connection with consciousness along with some possibility of being implemented in natural nervous systems. As it happens, there is a very strong candidate principle: the development and use of internal models.

There are of course a great many structures in information processing architectures that could be considered as being internal models of one sort or another. In addition, the number of different characteristics of the world, the robot, and their interaction that might be modelled — the physical, abstract, logical, temporal, and other aspects — is effectively unlimited. All we can do in the space available here is to sketch out some of the kinds of internal models that we believe might be useful within the context of this project, and in particular to identify their connections with control systems, intelligence, the brain, and consciousness. (It is difficult to find a good general theoretical treatment of this field, although it was a major topic of study within cybernetics; Arbib's *The Metaphorical Brain* [Arbib, 1972] is still more useful than many later texts.)

The sense in which the word 'model' is used in engineering is slightly different from its use in everyday speech. The basic characteristic of any model, internal or external, might seem to be that it is some sort of process or structure that in some way resembles whatever it purports to model. However, when the model is to be used for some purpose by the control system of a robot, the model will be represented within some internal information processing system, and the sole requirement is that information processing operations involving the model should yield *appropriate outputs* in relation to the aspects of the situation being modelled (Minsky, 1968). There is no intrinsic requirement for the model itself to correspond to reality in any other way. (Of course, this does not preclude the use of models which do have a clear resemblance to whatever is being modelled.) For example, in the field of artificial neural networks, a network that has been trained to produce a particular set of outputs in response to a given set of inputs is often said to have learned 'an internal model' of the problem; this internal model is simply a pattern of synaptic weights that happens to give the correct outputs, and it is only in rare cases or in certain special types of networks that the characteristics of the internal model can be related explicitly to the characteristics of the problem.

Engineers are often particularly interested in what might be called 'working models', in which induced changes in one part of the model cause all the linked parts of the model to change appropriately and in step with one another, so that the dynamics of the whole system is captured. This is different from having a database of facts, or a look-up table, where a suitably encoded enquiry produces the correct answer without the mediation of any active modelling process.

A potential problem with this approach is that the attempt to achieve intelligence in artificial systems by the formation and exploitation of internal models of the world has been going on in one form or another since the Dartmouth Conference of 1955; in fact, it is the programme of what is often referred to pejoratively as GOFAI (Good Old-Fashioned Artificial Intelligence). Its critics, notably Brooks (1991) and Dreyfus (1992), have claimed that the idea of controlling behaviour by building and reasoning over symbolic internal models (or representations) in this tradition is fundamentally flawed. We agree, but point out that other forms of internal models and other methods of exploitation are largely unaffected by their criticisms, and that the success of modelling approaches within engineering shows that complex systems can be controlled in real time by appropriate variants of these methods, as described below.

The Use of Internal Models in Control Systems

In engineering, the system that needs to be controlled — such as an aeroplane, oil refinery, or air conditioning unit — is known as the plant. The control system sends control inputs to the plant that affect the operation of the plant in various ways, and it receives outputs from the plant that it may use, along with environmental information, when calculating the control inputs. The purpose of the control system is to force the plant outputs to achieve or maintain some state, or to follow some desired trajectory. Environmental factors may also produce disturbances, creating the need for control actions. There are two broad classes of control: feedback, and feedforward.

In feedback control, also known as closed loop control, the control inputs are constantly modified as a function of the plant outputs. The simplest and best known form is negative feedback: the control task is to maintain some value of the plant output at a fixed level, and the controller generates a control input that that is a function of the difference, or error, between the measured value and the desired value, and tends to reduce that difference. If there is no error, the controller produces no output. Problems can arise when there are long time delays in the loop — for example, if the sensor measuring the controlled output value is very slow, or if the plant itself takes a long time to respond to a control input. Since the controller is driven by errors, it will not be able to achieve very rapid responses, and other techniques will have to be used.

Modern feedback controllers tend to use models of some kind to improve their performance. Rather than simply measuring the plant output that is to be controlled, measurements of many other variables can be taken so that the current state of the plant can be estimated accurately; this amounts to constructing a kind of model. In situations where the behaviour of the plant is incompletely known, or changes with time, a computer system running a more advanced controller (a so-called adaptive controller) may develop or adjust an internal model of the plant (a process known as system identification); alternatively, the controller may adjust itself so that the effect it has on the plant matches the computed effect on an idealised internal model of the plant. Adaptation is always a much slower process than the control process itself.

In feedforward control, sometimes called open loop control, the control inputs are calculated from the following factors: the desired control value; measurements of environmental variables (e.g. disturbances); and the current state of the plant. They are then passed to the plant, and are not modified by any plant outputs occurring after that time. Because these controllers do not have to wait for the sensed response of the system, they can operate extremely fast. A good feedforward controller will produce a control input that exactly cancels the effects of any environmental disturbance and achieves the target value for the controlled output. For example, a robot arm might have to be moved very quickly to a particular end point; a feedforward controller would calculate from the initial position and the desired position of the arm the exact timing and amount of electric current to be supplied to each joint motor to produce the required movement. This calculation can be thought of as requiring some sort of model of the arm's characteristics — an implicit model, expressed only in terms of the current patterns necessary to produce particular movements. This type of model is known as an inverse model, and is the opposite of a forward model, which yields the movements or sensory conditions that will result from the application of a particular configuration of joint motor currents.

More complex controllers, sometimes known as optimal controllers, or model-based multi-step predictive controllers, do not attempt to keep the instantaneous values of various plant outputs close to some desired profile, but instead operate to minimise the cumulative value over time of some cost measurement associated with the plant. They do this by 'planning' the sequence of control actions stretching into the future that is calculated to produce the lowest-cost sequence of plant changes; they then execute the first action, and calculate the optimal sequence afresh. In order to do this, they need a good model of the behaviour of the plant; again, there are techniques for acquiring or modifying such a model if it is partially unknown or if it changes over time. The most sophisticated controllers currently in use are usually called adaptive model-based predictive controllers.

The controllers mentioned above are usually implemented using conventional mathematical formulations based on linear systems, or on linear approximations to non-linear systems. The theory behind such methods is well developed, and makes it possible to design control systems with proven characteristics of stability and convergence. In the last fifteen years, some less conventional methods have been developed to cope with systems which are highly non-linear, or in which the nature of the system model is unknown. Some types of neural networks are able to learn accurate models of unknown non-linear functions, and have given rise to the field of neurocontrol (Werbos, 1990; 1992). Where a non-linear model is partly known, control systems using fuzzy logic offer rapid and accurate control (Passino and Yurkovich, 1998). The learning abilities of neural networks are sometimes combined with the tractability of fuzzy systems in neuro-fuzzy control systems (Nauck *et al.*,1997). These new methods are highly adaptable, and can be used as standalone systems, or as components of model-based predictive controllers or other techniques originally developed for linear systems.

It can therefore be seen that models of what is being controlled, and how it responds to control inputs, are widely used in control systems; these models are of various types, and are used for various purposes.

What Could a Robot or an Animal Use an Internal Model For?

We have identified four major applications which seem to be relevant for animals and robots, and which align well with our programme: processing novel or incomplete data; detecting anomalies; enabling and improving control; and informing decisions. As there is not the space to deal with them in any detail, we restrict ourselves to giving an example of each.

Processing novel or incomplete data. Any scheme that, in effect, allows a novel input to be mapped to a category, and treated as an exemplar of the category, can be regarded as involving a kind of model. In many animals, exposure to a stimulus with certain fixed characteristics (a 'sign stimulus') may release a stereotyped response; for example, the presentation of a red patch will cause a male robin to attack it, regardless of whether it is another robin's breast (the natural stimulus), a piece of cardboard, or even a Post Office van seen through a window. Regardless of the actual mechanism used, the robin (or other organism or robot) could be said to have some kind of implicit internal model of 'that-which-is-to-be-attacked'. Of course, if the mechanism involved the explicit comparison of a processed sensory input with some template, the template would correspond more closely to the everyday meaning of 'model'.

Detecting anomalies. If a model generates an expectation or prediction about what goes with what, or what follows what, then any unexpected conjunction or succession can be detected by the system of which the model is a part. This has survival value because novelty often indicates danger, or some other situation that is worth attending or responding to. It is in some ways the counterpart of the first use of models because it can be regarded as the category of unmodelled occurrences. For example, chimpanzees will tolerate a doll that looks like a normal chimp, but will respond with fear or aggression to a deformed or incomplete doll.

Enabling or improving control. Any of the examples given above in the context of model-based control will suffice.

Informing decisions. A model can be used in many ways to predict the consequences or utility of various possible future actions, and so can be used to guide the generation or selection of such actions. There are two basic scenarios for prediction: single step and multi-step. In a single step prediction, all that is predicted is the consequence of the next action; in a multi-step prediction, what is predicted is the consequence of a sequence of actions. However, a multi-step prediction does not necessarily imply the commitment to an entire sequence of actions, because it can also be used merely to select the first action from the preferred sequence, with the whole simulation and selection process being run after every executed action. This is typical of the approach used in engineering, and in many planning systems in AI. From a cognitive viewpoint, the outcome of any of these processes can be regarded as representing a decision; however, it should be remembered that the selection of one of a number of alternative possibilities can also be achieved by what some might see as a non-cognitive mechanism, such as a winner-take-all network. The role of the model is not to make the decision, but rather to provide information to the decision-making process.

The use of models to inform decisions is particularly relevant to our project, and most of what we want to convey is captured by Dennett in his description of a hypothetical creature, the last of three in an evolutionary sequence (Dennett, 1995). The first, the Darwinian creature, is the basic model. Its responses to its environment are specified by its genes; those examples with genes producing bad responses die, and those with genes for good responses survive to breed, eventually producing a population with better responses. The second, the Skinnerian creature, is capable of learning, and as a result becomes capable of producing better responses if it is not killed by an early bad response. The one we are interested in is the third, the Popperian creature, which is able to preselect its responses so that those likely to kill it are inhibited.

> But how is this preselection in Popperian agents to be done? Where is the feedback (about the quality of the proposed action) to come from? It must come from a sort of *inner environment* — an inner something-or-other that is structured in such a way that the surrogate actions it favours are more often than not the very actions the real world would also bless, if they were performed. In short, the inner environment,

> whatever it is, must contain lots of *information* about the outer environment and its regularities . . . we must be very careful not to think of this inner environment as simply a replica of the outer world, with all its physical contingencies reproduced. . . . The information about the world has to be there, but it also has to be structured in such a way that there is a nonmiraculous explanation of how it got there, how it is maintained, and how it actually achieves the preselective effects that are its *raison d'etre* (Dennett, 1995, pp. 375–6).

A more technical theoretical treatment of some of the ways in which techniques derived from control systems might be applied to cognition can be found in Grush (1997; 2002). Clark and Grush (1999) extend the analysis to the problem of achieving meaningful cognition in robots. Although the approach is primarily philosophical, the content could easily be adapted for implementation on a robot, and in that case would probably be very much in line with what is being proposed here.

The Presence and Use of Models in the Brain

There is a considerable amount of evidence that many of the control systems in the brain use models in much the same way as control systems designed by engineers. However, the brain also appears to exploit models in ways that go beyond the current capabilities of engineering systems. This brief review can do no more than scratch the surface of this topic, and the examples given below are a sample rather than a balanced view of an extensive literature; see Wolpert and Ghahramani (2000) for a more comprehensive presentation.

A major problem the brain faces it that its sensory systems are relatively slow — visual feedback takes around 100ms — and so feedback control cannot be used to control rapid movements. Since the brain does manage to produce appropriately timed and accurately modulated muscle activations to control rapid movements, it must be using a feedforward scheme with inverse models to do this. The variability of tasks, situations, and bodies means that these inverse models cannot all be innate; plausible schemes have been advanced concerning the ways in which they might be learned.

One characteristic of multi-jointed systems such as the human body is that there are an infinite number of ways of moving a limb to a specified position; the control problem is said to be underdetermined. However, there is a remarkable degree of stereotyping when an individual is asked to repeat a movement, and when different people are asked to make the same movement. This seems to be due to the brain using a form of optimal control, which selects from all the possible trajectories the one that minimises some cost function, and does so by using some kind of model. A lot of effort has gone into discovering the cost function that the brain uses; whatever it is, it tends to produce smooth movements with smooth changes of torque at the joints.

Brains also seem to use forward models. When given the motor signals supplied to the muscles, a forward model will yield the results, which may be expressed in terms of the final position of the limbs, or of the sensory inputs associated with that position. It has long been known that a copy — the efference copy — of the motor outputs from the brain is sent from the motor cortex to other parts of the brain. This appears to be used as the input to one or more forward models, which then yield the sensory input that the movement will produce. This can later be combined with the incoming sensory feedback, which will be noisy, to produce a more accurate estimate of the actual current state of the system. Perhaps more importantly, since the forward model computation can start as soon as the efference copy is received — which is before the movement starts — the output can serve as a prediction of the expected movement and sensory input. This prediction can be used in many ways. For example, the predicted movement can be compared to the intended movement, and any difference (arising from noise, or from a deficient inverse model) can be compensated for much faster (though less accurately) than if feedback had to be relied upon. In addition, it can enable the brain to distinguish between movements caused by voluntary action (where the actual sensory feedback will match the prediction), and movements caused or affected by external factors or forces (where there will be a mismatch); and it can also be used to cancel out any sensory disturbances caused by voluntary movements (such as head movements) that affect the inputs from sense organs. In many cases, the processing using these forward models can be localised and identified in the brain by modern techniques of brain imaging.

Outside the context of control, there are many indications that imagined situations may lead to brain activity similar enough to the brain activity caused by the real situations to be regarded as primitive models of them. Most of these studies involve sensory imagery alone (e.g. Behrmann, 2000) but some combine sensory imagery with some imagined action on the image — typically rotation — and find evidence of apparently appropriate activity in motor areas of the brain (e.g. Richter *et al.*, 2000). Hesslow (2002) has gathered together a great deal of relevant data in support of what he calls the 'simulation hypothesis'. He advances three 'core assumptions':

> (1) *Simulation of actions*. We can activate pre-motor areas in the frontal lobes in a way that resembles activity during a normal action but does not cause any overt movement.
>
> (2) *Simulation of perception*. Imagining that one perceives something is essentially the same as actually perceiving it, but the perceptual activity is generated by the brain itself rather than by external stimuli.
>
> (3) *Anticipation*. There are associative mechanisms that enable both behavioural and perceptual activity to elicit other perceptual activity in the sensory areas of the brain. Most importantly, a simulated action can elicit perceptual activity that resembles the activity *that would have occurred if the action had actually been performed* (Hesslow, 2002).

He is able to present some neuroscientific evidence in support of the first two assumptions. If evidence could be found to support the third, it would supply a mechanism to support Dennett's speculations about the use of models in intelligent behaviour, because, as Hesslow notes, 'Once the mechanism of anticipation is in place, there is nothing to prevent the appearance of long chains of simulated responses and perceptions.'

The Possible Relationships Between Internal Models and Consciousness

There are many indications of the possible involvement of models in consciousness. The first is that conscious experience, especially when problem solving, sometimes seems to involve some sort of model of a real-world object — for example, when imagining a room in your house to see if a piece of furniture in a shop will fit. These 'mental models' have particular characteristics, and both these and the ways in which they can be used to solve problems have been studied in some depth by psychologists such as Johnson-Laird (1983). Johnson-Laird takes his inspiration from Craik, who wrote (at a time when intelligence and consciousness were regarded as inseparable):

> the nervous system is . . . a calculating machine capable of modelling or paralleling external events' (Craik, 1943, p. 120);

> If the organism carries a 'small-scale model' of external reality and of its own possible actions within its head, it is able to try out various alternatives, conclude which is the best of them, react to future situations before they arise, utilise the knowledge of past events in dealing with the future, and in every way to react in a much fuller, safer and more competent manner to the emergencies which face it (Craik, 1943, p. 61).

A second indication of the involvement of models comes from the spectacular examples of internal modelling recently made available by Ramachandran and his collaborators (Ramachandran and Blakeslee, 1998). They involve so-called 'phantom limbs' — the internal image-like experienced representations of limbs that are absent, either through amputation or through developmental failure. In some cases, phantom limbs give rise to chronic pain or discomfort because the individual has lost the capacity to 'move' a phantom limb, and it has ended up in an awkward position. By using cunning experimental arrangements to replace the visual feedback that used to come from the now-absent limb with stimulation coming from a mirror-image of the remaining limb, Ramachandran has enabled some patients to regain their control of the phantom limb, and to move it to a comfortable position. This shows that the internal model is in some way affected by input from the real limb. On the other hand, a few rare cases reveal that phantom limbs can be present even in the congenital absence of limbs, showing that the model may be innate rather than learned. It is often difficult to make sense of these reports, which vary considerably, but the strong conscious awareness of one or more internal models of the disposition of the limbs seems beyond doubt.

A third line of thought suggests that the propensity of the brain for modelling may lead directly to consciousness itself, rather than merely providing the 'contents' of consciousness. In 1976 Dawkins set out his view that animals can be regarded as machines designed to ensure the survival and propagation of their genes, and remarked:

> Survival machines that can simulate the future are one jump ahead of survival machines who can only learn in the basis of overt trial and error. . . . The evolution of the capacity to simulate seems to have culminated in subjective consciousness. . . . Perhaps consciousness arises when the brain's simulation of the world becomes so complete that it must include a model of itself (Dawkins, 1976).

This intriguing speculation seems to have been rather neglected; it is occasionally mentioned in the literature, but there does not seem to be any body of work directly inspired by it. Earlier, Minsky (1968) had identified some of the problems of thinking about consciousness as being caused by attempts at 'explaining the complicated interactions between the parts of the self-model', and had discussed systems containing both world-models and self-models, but had not proposed that the possession of a self-model might in itself give rise to consciousness. Perlis, in a series of papers summarised in Perlis (1997), comes closer to Dawkins with his '*amazing-structures-and-processes paradigm*', in which a self-modelling process enabling what he calls 'strong-self-reference' somehow gives rise to conscious experience. His work is particularly interesting in the present context because he explicitly considers the possible role of robots in shedding light on consciousness:

> I think that only by careful examination of human behaviour and the design of smarter robots will we be able to position ourselves to have more than merely prejudicial intuitions. . . . No-one, to my knowledge, has built, or even tried to build, strongly-self-referring machines. This in large part is simply due to the fact that no one has tried to build robots that can do very much reasoning, or even that can do very much common-sensical self-protection in a complex world. But strong self-reference is what an intelligent robot needs . . . (Perlis, 1997).

More recently, Damasio (1999) has proposed a neurologically based theory of consciousness in which the development of a primitive body-centred self structure plays a crucial role. The theory itself is complex, but his hypothesis is well summarised by Churchland (2002) in a paper examining self-representation in nervous systems:

> . . . the self/nonself distinction, though originally designed to support coherencing, is ultimately responsible for consciousness. According to this view, a brain whose wiring enables it to distinguish between inner-world representations and outer-world representations and to build a metarepresentational model of the relation between outer and inner entities is a brain enjoying some degree of consciousness. . . . Conceivably, as wiring modifications enable increasingly sophisticated simulation and deliberation, the self-representational apparatus becomes correspondingly more elaborate, and therewith the self/not-self apparatus. On this hypothesis, the degrees or levels of conscious awareness are upgraded in tandem with the self-representational upgrades (Churchland, 2002, p. 310).

The important point here is that it is not just the existence of some kind of self-representation that is held to lead to consciousness — it is the self-representation's relation to representations of things in the world, an idea that is more specific than Dawkins'.

An Incremental Programme

Our plan is to build a succession of robots that deal with the world by building and exploiting internal models; each new robot will be derived from, and capable of dealing with the world more intelligently than, its predecessor. Broadly speaking, there are two possible methods of doing this. The first would begin with a simple robot, embedded in a human-like world, and would attempt to increase its ability to cope with this world step by step; the second would move in stages from simple environments through progressively more challenging ones, adding the necessary extra abilities at each stage. Brooks (1991) has stressed the virtues of the incremental development of robots in the final target environment, and the pitfalls of beginning work in artificially simpler situations and hoping to migrate to more complex worlds. While we endorse his comments, we believe that the second strategy is more appropriate for this project; the reason is that we need to understand the situations and contingencies for which modelling may deliver benefits, and so we need to be able to control the exposure of the robot to those situations and contingencies. Our primary concern is not merely with building a robot that can cope, but also in understanding how modelling can enable it to cope; we therefore need to control what is present to be modelled, and what contingencies enable what kinds of modelling to deliver benefits. However, we recognise that the risks pointed out by Brooks may be inherent in this choice, and that a system able to cope with a simple situation may be wholly unable to cope when the situation is made even slightly more complex.

The path we propose to take is to begin with a very simple robot with a simple lifestyle (or mission) in a simple — but real — world. We will give it the ability to cope with that world, and will then progressively increase the complexity of its world and its lifestyle, and give it the bodily and computational resources to deal with the new problems. This is close to recapitulating evolution, but we do not commit ourselves to using some form of artificial evolution as the only mechanism, or even at all. Design is fast, comprehensible, modifiable, and extensible, and is what engineers are good at. Evolution is slow, intractable for most physical systems, and is not guaranteed to be comprehensible — indeed, there would be little point in evolving consciousness a second time, because the likelihood is that we would then be faced with a second type of system manifesting a consciousness that we could not guarantee to understand. In pursuit of the goal of producing a system that can be understood, we must at all times ensure that our systems are as *transparent* as possible — that we can see how they work, and in particular that we can see what their internal models are like, and how they deliver benefits.

Why use a real robot when using a simulated robot would be faster and more convenient? Parts of this question were answered very well by Brooks in a series of papers, including Brooks (1991). For example, using a real robot forces the use of the real world, and that prevents the avoidance, deliberate or unintentional, of the difficult problems inherent in the constraints of real world physics and real time. As far as convenience goes, it might well be easier to simulate a robot than to build it, but, as Brooks remarked, it would be much more difficult to simulate the rest of the world than simply to use it. Perhaps most importantly from the perspective of this project, our human consciousness developed as a result of evolution in the real world, and so if we were to succeed in developing some consciousness-like phenomena in a real robot, it would be reasonable to compare it with our own consciousness. If a modelled entity in a modelled world showed some consciousness-like phenomena, a critic could argue that any connection with our own consciousness would be formal or analogical rather than real: the consciousness of a simulated entity could only ever be a simulated consciousness. However, these considerations do not necessarily rule out the use of simulation in situations where the robot and environment are simple enough to give confidence that the simulation will be adequate, or when all that is required is an indication of whether a particular path is worth pursuing or not.

Many aspects of our programme are similar to those being followed by Edelman in his series of experiments with the Darwin robots. Edelman and his collaborators are attempting to understand the operation of the mammalian brain (Edelman, 1987) by using progressively more complete simulations of the brain to control a succession of progressively more complex robots in progressively more complex environments (e.g. Krichmar and Edelman, 2002). Our aim is to understand how the use of modelling to increase the capacity for intelligent behaviour may produce phenomena similar to those associated with consciousness. However great the difference in aims, in practice we will be using very similar equipment (robots, digital computers) and a very similar incremental strategy, and much of what he has said about the rationale and conduct of his decade-old programme will also apply to ours.

The Starting Point

The obvious way to start our programme was to place a simple robotic system in a simple environment, and to implement a simple modelling scheme to demonstrate the four uses of internal models. The aim was to create a system with the key fundamental attributes of any control scheme based on internal models; this could then serve as a baseline level of both structure and performance, from which we could pursue our strategy of incremental development. We did not expect to learn anything about consciousness from constructing such a system, and we were sure that its simplicity would deter others from over-interpreting any aspects of the system or its behaviour.

We began by looking for a suitable modelling scheme. An early candidate was the neural network implementation of a dynamical systems approach developed by Tani (1998). This was all the more attractive because it formed the basis of what we believe to be the first serious investigation of consciousness-related emergent phenomena using a robot as an investigative tool. For example, one of Tani's conclusions was that 'There is an essential structure of the "self" in the system and occurrences of "self-consciousness" are explained in terms of unfolding of this structure in time' (Tani, 1998). However, we felt that Tani's system was more complex than necessary for our purposes, and did not offer sufficient transparency, so we eventually settled on ARAVQ (Adaptive Resource Allocating Vector Quantization), a technique developed by Linåker and Niklasson (2000a,b) for dealing with abstract sensory flow representations.

For many reasons, it is useful for a robot to store sequences of past sensory inputs, or 'experiences', but memory limitations usually mean that relatively few experiences can be stored; compressing the data enables more instances to be stored, but not in a form that is immediately useful. Linåker and Niklasson attacked these problems by designing a method that allowed 'sensory flows' — raw temporal sequences of sensor inputs — to be encoded economically in a *useful* form. They had observed that when a simple robot is carrying out a simple task in a typically simple environment, the robot's sensory inputs (and also its motor outputs) are in practice often relatively stable for long periods of time. For example, when a robot is following a wall (a typical and much studied low-level robotic task), both its distance from the wall and its speed will be fairly constant, and so the sensor inputs from the robot's rangefinders, and of course its motor output commands, will also tend to be constant. If the robot then enters a narrow corridor, a typical control program will cause it to follow a course central to the corridor, often at a reduced speed; again, the sensor inputs (from both sides of the robot) and the motor output commands will tend to be constant, but different from those in the previous wall-following situation.

Using a simple robot simulation, Linåker and Niklasson began by regularly sampling each input configuration — the instantaneous sensory input and motor output — and developed a simple on-line clustering algorithm, ARAVQ, to detect and describe the small number of relatively stable and distinct input configurations, which they called 'concepts'.[1] Their robot could store long sequences of experiences very economically simply by labelling the different concepts, and recording the number of times each concept was repeated consecutively. For example, a run in which the robot followed a wall (concept W) for 29 time intervals, rounded a slow bend (concept B) for 4 time intervals, and then moved along a corridor (concept C) for 37 time intervals would reduce to (W,29; B,4; C,37).

The concepts have some interesting properties. Each one represents a range of input configurations which the algorithm has identified as forming a cluster. If a new configuration falls within that range, or close enough to it, it will be mapped to that concept, and the concept itself may be slightly modified to represent the new data rather better. However, if the new configuration falls far enough outside any existing concept, it may be used as the basis of a new concept; if more configurations close to the new concept are encountered, they will then be assimilated to it. We refer the reader to their papers (Linåker and Niklasson, 2000a,b) for a detailed description of their algorithm.

An attractive feature of this scheme is that it is possible to take a particular sequence of concepts recorded by the robot, and to use it to construct a step-by-

[1] We are unhappy with their terminology, but we retain it here for compatibility with their work.

step representation of the robot's 'experience' and 'behaviour'. Each distinct concept can be characterised by the average reading of each of the sensor inputs and motor outputs, and so for each occurrence of a concept, it is possible to construct an approximate representation of (a) what must have been in the environment to activate the rangefinders, and (b) how the robot must have moved during the sample period. By plotting these movements step by step, and at the same time tracing out the points corresponding to the rangefinding data — a process Linåker and Niklasson call 'inversion' — it is possible to produce a representation of the robot's implicit model of its movement through the environment, and of its sensory contact with environmental features. The result looks like a map of the environment, and enables a quick appreciation of how well the implicit model corresponds to the real environment. Examples of the results of inversion are shown and discussed below.

Linåker and Niklasson used a simulation of the popular Khepera robot (Cyberbotics, 2003b). The basic Khepera, although it is a complex and beautifully engineered product, is conceptually very simple: it is circular, with two independently driven wheels on each end of a diameter, and with six equally spaced rangefinders mounted on one semicircumference (facing forwards) and two on the other, facing backwards. Our first step was to reimplement their work using Webots, the very sophisticated and accurate simulation system provided by the Khepera's manufacturers (Cyberbotics, 2003a,b). Figure 1 shows a screen shot from the simulation of our first experimental environment. The simulated Khepera is programmed with a simple wall-following routine, and moves forward keeping a set distance from the wall on its right while avoiding collisions. We found that ARAVQ rapidly built a stable set of concepts, and that it was easy to identify these concepts with particular sections and features of the environment.

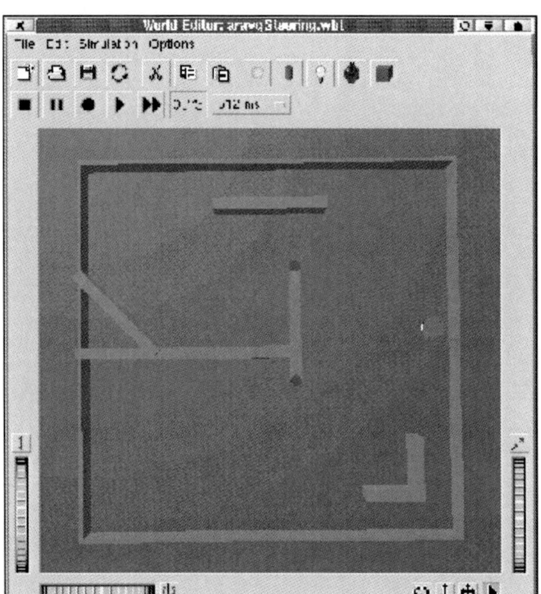

Figure 1. A screenshot of the environment and Khepera modelled in the Webots simulator. (Copyright CyberBotics Sarl)

In Figure 2 we show how the process of inversion operates. The image on the left represents the inversion of two successive activations of the same concept from a Khepera moving along a corridor. (The size of the step between the activations has been scaled up for clarity.) The Khepera is represented by a circle, with the front-back and left-right diameters drawn in, and with a triangle superimposed on the left-right diameter to show the orientation. The positions of the rangefinders are

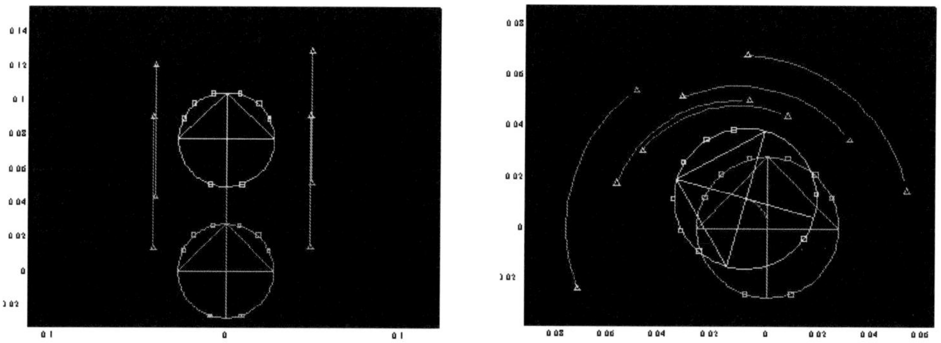

Figure 2. How a map of the environment can be reconstructed from the record of the robot's concept sequence.

In the record on the left, the 'corridor' concept is active for two successive time intervals. The model of the Khepera is displaced by an amount corresponding to the concept's motor activation. The sensor readings for the two leftmost and two rightmost sensors give the distance of the reflecting surfaces (walls) at each position; by joining successive corresponding points, we can see part of an environmental map. The record on the right shows a transition involving a small translation and a large rotation, with the derived map. (After Linåker and Niklasson, 2000a)

indicated by the small circles on the periphery of the robot. In the initial (lower) position, small triangles mark the points corresponding to the readings of the two leftmost and the two rightmost rangefinders, as encoded in the concept. In the upper position, the representation of the Khepera has been moved by an amount corresponding to the operation of the left and right motor outputs for one time step; since the outputs encoded in the concept are equal, the Khepera has simply been moved forwards. The new points defined by the rangefinder readings are marked, and lines are drawn joining the successive readings of the points defined by each rangefinder. These show the line of the corridor quite accurately.

The image on the right shows two successive activations of a concept formed by a Khepera turning in response to a corner. In the initial position, all the front-mounted rangefinders except the leftmost detect the right-hand wall of the corner; its approximate position can be seen by joining the rightmost triangles on each of the five curves traced out by the sensors. The Khepera's motion, as encoded in the concept and as illustrated by the change in position and orientation of the triangle on the diameter, is a left turn of almost 90° combined with a small forward movement. Since the concept is the same as at the previous step, the relative positions of the rangefinder distances are the same in relation to the Khepera; as can be seen by joining the leftmost triangles on each curve, they mark the approximate position of the other wall of the corner. Although the line of the corner is clear enough once one is used to the representation, it is less intuitive than the image corresponding to a wall or corridor.

In Figure 3 we can see the results of inverting the series of concepts activated by the simulated Khepera in making a circuit of the environment of Figure 1. The local representation of detail is surprisingly good, but the global map is distorted, primarily because of errors associated with rotational movement — a well

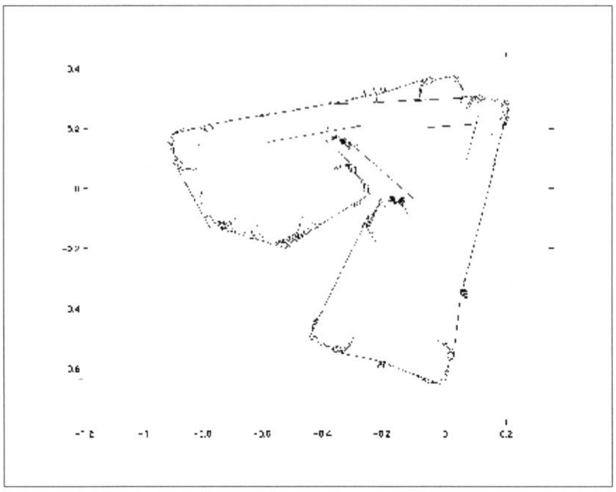

Figure 3. The environmental map derived from a simulated Khepera in the environment of Figure 1.

Although rotational inaccuracies distort the map, it contains good local representations.

known problem in robotics. (Robot simulations are often criticised because they are held to give insufficient weight to the real-world noise and random errors which affect the sensors and motors of real robots. In Webots, noise can be added to the simulated inputs and outputs to compensate for this.)

In order to bring this work into the real world, our next step was to see if ARAVQ was capable of forming stable and potentially useful concepts in a real Khepera operating in a real environment. Perhaps because of the Khepera's lack of dynamics and simple sensory apparatus, we encountered no problems in making this transition. Figure 4 shows a typical environment being explored by a real Khepera programmed with the usual simple wall-following algorithm; Figure 5 shows the results of inverting the sequence of concept activations for a single circuit once the concepts had stabilised. As can be seen, the representation of the implicit model appears qualitatively similar to the results obtained in simulation, with the good local detail, but with the global representation distorted by rotational inaccuracies.

We have so far shown that a robot using the modelling scheme, ARAVQ, can build an internal model of its interaction with an environment, and that this internal model (comprising both the concepts and the list of the sequence of concepts encountered in the environment) can be interpreted to produce a comprehensible representation of the outcome of the interaction. We now need to show that the model can support the four functions of modelling identified above; however, it will be convenient to examine them in a different order.

Enabling and improving control

The Khepera had learned the concepts passively, while moving through the environment under the control of the wall-following program. In order to make the robot move under the active control of the concepts, it was clear that the basic requirement would be to select a concept corresponding to the existing stimulus configuration, and then execute the motor outputs represented by that concept. However, the motor outputs were themselves part of the concept — how could they be taken into account before they had been generated? Nathan Grey, who carried out the robot experiments, suggested an elegantly simple method: select the most appropriate concept solely on the basis of the rangefinder data, and then execute the motor outputs specified by that concept. The results at first surprised

Figure 4. The Khepera robot is modelling the environment using ARAVQ.

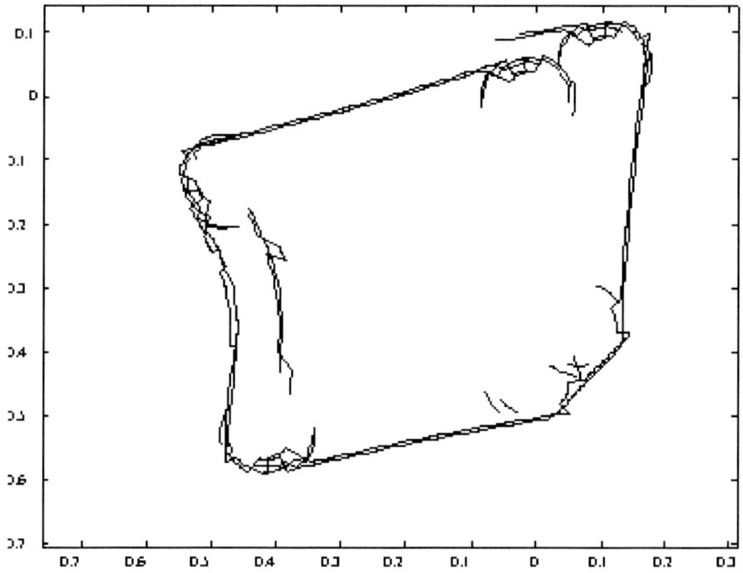

Figure 5 The environmental map derived from the modelling process of Figure 4.
Although the global map is distorted, the local detail is excellent.

us. Under the control of the concepts, the robot produced smooth and accurate wall-following — a great improvement on that produced by the original wall-following program! On reflection, however, this was to be expected. The wall-following program had been made deliberately crude and jerky, responding to inputs with large changes in motor output, because we wanted to present ARAVQ with a reasonably difficult modelling problem. Since the motor output specified by a concept is an average of the motor commands that have been produced within that concept, it is bound to vary less than the outputs that contributed to the formation of the concept, and so local smoothing is inevitable.

We had thus used one facet of the internal model — the concept — to enable control; by accident, it had also improved control. It should be noted that we disabled the concept formation during this test phase.

Processing novel or incomplete data through an existing model to produce 'appropriate' actions

When the robot is run in the mode described above, the incoming set of rangefinder readings will simply be mapped to the concept with the most similar set, as defined by a simple measure. Since there are six rangefinders, each taking readings independently, it is unlikely that exactly the same set of readings will be encountered twice in a single run; because of minute environmental variations, this would be true even if the rangefinders behaved perfectly, but the certainty of sensor error (especially occasional failed readings) makes it something to be expected anyway. Because of the simplicity and regularity of the environment, and the relatively weak demands of the control task (wall-following), the selection of the concept with the most similar input set did in fact produce appropriate actions from novel or incomplete data.

Detecting anomalies using a model

There are two distinct ways in which ARAVQ could be used to detect anomalies. If a persistent and stable input configuration is not close enough to an existing concept, ARAVQ automatically creates a new concept; this step could be used to signal that an anomaly had been detected. (By design, ARAVQ effectively ignores momentary or unstable variations from familiar inputs, since these situations occur very frequently in robotics because of environmental variation and sensor noise.) In addition, if the sequence of concepts encountered when traversing a familiar environment did not match the stored sequence (again, within limits), this could be used to signal that something had changed. Because of their simplicity, we did not think it necessary to test either of these scenarios.

Using a model to inform decisions

Since the wall-following task (following a wall on the right) produced fairly invariant linear paths around the environment, there was not a lot of scope for presenting the robot with a decision to be made on the basis of its internal models, because it could do nothing about choosing one route over another. However, it could at least decide whether to stop or continue, and so we took inspiration from the behaviour of a prey animal on detecting a predator. If the animal is close enough to home to make a successful run for it, it will do so; if it is too far away, it may freeze into immobility, and so avoid detection. We played the part of the predator by giving the robot a signal while it patrolled the periphery of the environment. The robot's 'home' was defined as a short corridor section with a distinct concept corresponding to it, as can be seen in Figures 4 and 5. On receipt of the signal, the robot internally stepped through the stored sequence of concepts, one by one, from its current location. If the home concept was encountered before some internal time limit elapsed (corresponding to internally stepping through about half the journey round the environment) the robot would continue moving; if it was not, it would halt until given a signal to proceed. This was trivial to program, and of course it worked.

We have thus succeeded in establishing a baseline for our project — a robot that uses perhaps the simplest of internal models to achieve the four functions we have identified as being of interest. We have also made the internal models transparent to some degree. There is, we hope, no sign of anything to which anyone might want to apply the adjective 'conscious', although the system may perhaps qualify as being 'cognitive' on some definitions (Clark and Grush, 1999). There is nothing that we would wish to characterise as any sort of self-model, although the details of the robot's embodiment are clearly crucial in determining the form of the representations of the internal models. The only way is up.

Increasing the Complexity of the Robot

The nature of the robot's sensors obviously places some limitations on the nature and extent of the information the robot can acquire about the environment and its contents. The same is true of animals, as von Uexkull recognised many years ago (von Uexkull, 1934). A robot fitted only with horizontal range sensors would never be able to discriminate between a low structure and a high one with the same size and shape at the base, however important the distinction might be for its well being or survival. On the other hand, an extra set of range sensors mounted high on the robot would be utterly useless if the environment contained nothing that was able to stimulate them. Somewhere in between is the case where the robot can acquire information about some aspect of the environment, but there is no contingency where that information can make any difference to the robot's performance. In a given environment, therefore, the sensory provision should enable information to be extracted and used (perhaps in a modelling process) to provide benefits at least equalling the cost of that extraction and use. It will not be necessary to provide an improved capacity for sensing every time the environment is changed; the vertebrate eye has remained essentially the same for a very long time — it is the capacity of the brain for processing the information from the eye that has changed.

As regards effectors, it is worth noting that the majority of research robots can do nothing except move through the environment. This is not solely due to the lack of any grasping or manipulating appendages, because many robots are potentially able to act on their environment and change it without any specialised effectors, simply by pushing things; the problem is simply that the experimental environments rarely contain anything that can be pushed! Actions more complex than pushing require some specialised effectors, and these in turn require the presence in the environment of suitable structures or objects upon which they can act. In the context of our robots, or of animals, there is a further constraint very similar to that noted for the sensory equipment: acting on the environment should be able to bring some benefit, otherwise the provision of the effectors represents nothing more than a cost. Of course, effectors, especially if they have redundant degrees of freedom, may require modelling for control purposes. In addition, effectors enabling the manipulation of an object can yield information about the object that would not be obtainable in any other way — for example, how to manipulate the object, as well as how the object can be manipulated. Finally, any addition to an agent's complement of effectors is likely to increase its potential for engaging the environment in different ways, and this will enlarge the set of affordances offered by the various environmental features, again adding to the complexity of the system as a whole.

The state of a typical mobile robot such as the basic version of the Khepera is described by its pose alone — its position and orientation — simply because the body is rigid. However, once there are movable body parts or effectors, especially if they are articulated, their dispositions may need to be taken into account by the control system. If these body parts or effectors carry or serve any sensory apparatus, then their movements and positions may also be useful to the sensory system; some examples of the use of such information were given in the section above on the use of models in the brain.

The three factors mentioned so far — sensors, effectors, and the body — would serve to define the non-behavioural complexity of most present-day robots, the general idea being that the complexity of a robot will increase with increasing complexity in its sensors, effectors, and body. However, these factors would be inadequate to describe most animals, because there is a further determinant of complexity to take into account. Since animals are biological entities, they are subject to the biological constraints of metabolism, growth, and ageing, and these complicate matters. The most appropriate current action for an animal depends on the state of variables such as hunger, thirst, fatigue, on whether it is a juvenile or an adult, and so on; the only corresponding factor in most robots is the state of their battery, though robots with artificial metabolisms are beginning to appear (Kelly *et al.*, 1999; Wilkinson, 2000). Many of the shifting multiple goals that animals have to deal with come from these biological characteristics, and we think it likely that a robot with the animal-like intelligence that might lead to consciousness would have to be capable of dealing with similar time-varying, cyclic, and historical constraints.

Increasing the Complexity of the Environment

There are many aspects to a robot's or animal's environment. It is both a space, and a range of objects and materials contained and localised within that space. The techniques used by animals to move through space, and the extent to which they use modelling, are becoming increasingly well understood. Paradoxically, some of the most difficult environments spatially are those that appear least complex, such as the Sahara desert; environments rich in spatial information are relatively easy for both animals and robots to deal with. What makes an environment challenging and complex is the nature and disposition of its contents. If we take the baseline environment as being the simple fixed and bounded space of the baseline experiment, it is possible to see how the introduction of various types of objects will require the robot to improve its abilities, especially where modelling is concerned.

The first step might be the introduction of moving objects. These will force the robot to avoid them in real time; distal sensors will be necessary for this, and in a cluttered environment the ability to predict an object's trajectory may prove necessary. The Khepera's sensors are inadequate for this task, and so it is likely that the next set of experiments will require a new robot fitted with long-range ultrasonic or laser rangefinders, or a suitable vision system. The next step could be the introduction of a variety of objects with different values for the robot, perhaps corresponding to food values and levels of toxicity; the relationship between the sensory characteristics of the objects and their value will have to be learned. (This is the stage at which Edelman is currently working — see Krichmar and Edelman, 2002). If objects of different types tend to occur in particular parts of the environment under particular circumstances, advantage will be gained by modelling and exploiting this. If the value of an object can only be obtained by a certain sequence of actions (as in cracking a nut), rather than by simply gripping it as in Edelman's scheme, the problem posed by the environment becomes even more difficult.

Passive objects such as sticks, stones, fruit, and plants are clearly important to many animals, and to humans, but the really challenging components of most animals' environments are active objects, or agents. These may be of other species, and may be parasites, prey, predators, and so on. In dealing with any of these agents, reactive strategies will inevitably be outperformed by predictive strategies, and these will require some form of modelling. However, the most interesting agents to introduce into the robot's environment will be robotic analogues of conspecifics, which may be competitors, collaborators, or mates, or even all three at different times. Where there are repeated contacts with the same individuals, and where there are individual differences, it may become necessary to be able to identify individuals, and to remember their characteristics and the outcomes of previous interactions. Of course, we will now be at the level at which various theories have proposed the origination of social intelligence, communication, language, and consciousness; the important thing is that we will have reached it in an incremental way, and will have a complete knowledge and a good understanding of the modelling and other resources available to the robot enabling it to cope with the environment before each new introduction.

Effective Choice and Planning: Could, Should and Would

Although we distinguished four possible roles for modelling in a previous section, it is clear that one, the use of modelling to inform decisions, is more complex than the rest, and capable of much greater development. This is not the place for a full analysis of the system features and capabilities required to implement and support such a scheme; indeed, to the best of our knowledge, there is as yet no such analysis in the literature. However, some aspects of what must be capable of being done are particularly relevant to the development of consciousness, and we describe them briefly below. They can easily be organised under three headings: what the system *could* do, what it *should* do, and what it *would* do.

*What the system **could** do*

For effective planning, the internal models used by the system must reflect the real world with sufficient accuracy. There are two important aspects to this. First, the modelled properties of the physical world — objects and space — must reflect the real properties sufficiently accurately. Second, the abilities of the system to act on and in the physical world must also be modelled well enough, and this means modelling the physical properties of the body, the ability to control it, and also the ability to process information appropriately. As Ramachandran and Blakeslee note:

> [I]t is always obvious to you that there are some things you can do and others you cannot given the constraints of your body and of the external world. (You know you can't lift a truck . . .) Somewhere in your brain there are representations of all these possibilities, and the systems that plan commands . . . need to be aware of this distinction between things they can and cannot command you to do. . . . To achieve all this, I need to have in my brain not only a representation of the world and various objects in it but also a representation of myself, including my own body within that representation. . . . In addition, the representation of the external object has to interact with my self-representation . . . (Ramachandran and Blakeslee, 1998, p. 249).

Ramachandran's and Blakeslee's comment about the interaction between the self- and the object-representation is particularly interesting, and worth exploring more deeply. If a system is modelling two objects interacting — for example, two billiard balls colliding — then one obvious way to organise this is to allow separate instances of models of each object to interact, rather than to have a dedicated integrated model of the two objects interacting under all possible circumstances. (Incidentally, there is plenty of evidence that humans and animals are supplied with innate knowledge of a wide range of object properties, including their behaviour in collisions — e.g. see Hauser, 2000.) Although the most obvious models available to the system would be those based on perceptual information, as suggested by Hesslow (2002), any other adequate representation would do. When the situation being modelled is the body interacting with an object, the same approach of allowing models to interact seems appropriate, especially as Ramachandran and Blakeslee present evidence for an innate model of at least some characteristics of the physical body, namely the limbs. However, there is a difference between a model of the body and a model of an object such as a billiard ball: the interactions of a body model with an object model as do not just depend on the modelled physical properties of the two entities, but depend also on how the body model is controlled, and this in turn depends on the information available for controlling the body model. Modelling the catching of a ball requires the modelling of the trajectory of the ball, the modelling of the perception of the trajectory, and the modelling of the arm and hand movements produced by control systems using data from that perception. This suggests that object models based on perceptual data might be particularly useful for body–object interactions.

In this context, it may be worth pointing out that the development of virtual reality technologies has provided a powerful metaphor for many aspects of consciousness, one that is used by Metzinger (2000) and Dawkins (1998), among others. Virtual reality provides an individual with sensory inputs — chiefly vision, touch, and sound — which match the inputs that would be produced by the physical environment being modelled *if the individual were at a particular location* in that environment. The physics of the environment — the positions and masses of objects, the way balls bounce, the way light behaves — may be modelled and used in the underlying computations, but what is presented to the individual is not a three dimensional physical model of the environment in x, y, and z coordinates, but a simulacrum of the view from a point in such a model. The visual imagination also seems to use such a centred view, and this may indicate that at least some of the models in the brain are similar to the products of virtual reality. However, Grush (2002) has clearly set out the differences between purely sensor-based modelling (modal emulation, in his terminology) and modelling which deals with what he calls the *egocentric space/object environment* (amodal emulation); as he makes clear, visual imagery may well involve both types of systems.

What the system ***should*** *do*

If evolution has done its job perfectly, then out of all the actions or sequences of actions that an animal *could* perform, the one it *should* perform will be the one that can be expected to make the highest marginal contribution to the propagation of copies of its genes. There is clearly no credible way for any animal to calculate this exactly at every instant; some approximate computation, or implicit computation, must be carried out instead. In some cases, however, the result of such computations can be uncannily accurate. For example, when life's problems are simplified by a specific situation — feeding young — so that maximising the rate of acquisition of energy from food is all that matters, some birds can make decisions about when to gather food at particular time-dependent food sources (tidal mud, rubbish dumps) that appear to be optimal.

Methods for the generation and evaluation of plans, by algorithmic and heuristic means, have been at the centre of conventional AI since the discipline was founded. Unfortunately, there is little or no useful information about either the representation of sequences of actions in the brain, or the mechanisms by which sequences of high utility are generated rather than those of low utility. However, there is a wealth of evidence that the encoding of utility in the brain is strongly associated with emotion. Rolls (1999) examines the issue at length, and suggests that the notions of reward and punishment provide a unifying structure — a common currency — with plausible links to the mechanisms of choice and multi-step planning. 'Is there any alternative to such a reward/punishment based system in this evolution-by-natural-selection situation? I am not clear that there is' (Rolls, 1999, p. 273). Edelman is also convinced of the necessity for such a value system, and again chooses a scheme with a positive and a negative aspect based on appetitive and aversive events (Sporns *et al.*, 2000). The implication of such schemes is that the action sequences generated by an animal are those that either maximise the expected reward, minimise the expected punishment, or produce the most positive net effect. The usefulness of such decision-making is determined by the correspondence between those decisions and those which would be made by some ideal system using values derived from the actual expected reproductive benefits. We propose to adopt such a system in our project.

Of course, we do not mean to imply that there is nothing more to emotion than the production of a reward or punishment signal. Strong emotions may trigger behaviour such as fighting or flight, or cause bodily changes preparing the organism for such actions. In social species, the expression of emotion plays a key role in social interaction. Human consciousness is saturated with feelings and emotions varying on many more dimensions than mere sign and strength; we do not know why this is so, but its reality is unquestionable. However, from a functional point of view, we need some method for the comparative evaluation of action sequences in order to achieve the selection of the most advantageous, and the assessment of each sequence and each component action along a single positive–negative dimension is both simple and apparently biologically plausible.

The assumption made explicitly by ourselves, and sometimes implicitly by others, is that the generation of action sequences of high utility involves the sequential modelling of possible actions and of their objective effects, together with some evaluation of their benefit to the organism. Somehow, actions or action sequences evaluated highly tend to be executed, and good enough sequences are executed often enough to give some advantage over systems without this capacity. There may be some method of achieving all this without modelling, but we cannot at present imagine it. However, there is no point at all in modelling unless there is an evaluative component; unless the process leads to the agent doing what it *should* do, modelling will confer no benefit.

The idea that sequential modelling is at the root of planning was central to the early efforts in artificial intelligence. It soon became apparent that, in the paradigms used at the time, the key problem was not just to calculate the outcome of a particular sequence of actions, but to search among a number of possible sequences to find one that was acceptable. AI soon became centred around the exploration and evaluation of different search methods. It may well turn out that issues related to search will be important for the ideas presented here, but our immediate concern is with the nature of the sequential modelling itself.

What the system ***would*** *do*

Even if a system can identify the actions it is possible for it to carry out in various situations, and is able to run internal models of multi-step plans, and to estimate the rewards, punishments, and costs of each stage in the plan, that may not be the end of the story. Rolls believes that the human brain contains two independent methods of selecting action: the implicit system, also present in non-human primates, which selects the next action on the basis of 'assessment of the reinforcement-related value of a stimulus' (Rolls, 1999, p. 256); and the explicit system, a language-based long term planning system. He remarks of the hypothesised explicit planning system: '*The process may enable an available reward to be deferred for another reward that a particular multistep strategy could lead to*' (Rolls, 1999, p. 270). The implicit system is essentially for selecting the immediate action with the highest expected reward; the planning system will only confer benefit when it selects an action with a lower immediate reward (as judged by the implicit system) that will lead to an eventual higher net benefit. If Rolls' conjecture about 'dual routes to action' (Rolls, 1999, pp. 255–61) is correct, then for the planning system to operate successfully, it must in some way resolve any conflict with the implicit system in its own favour. It seems unlikely that this will be achieved by the planning system always completely suppressing the older and faster implicit system; we might expect to see a degree of flexibility, with the implicit system being more likely to win if the immediate reward is high and if the eventual reward favoured by the explicit system is remote or only slightly superior.

If this is the case, then it adds a new dimension to the formulation of a plan. To be effective, the planning system must be able to predict whether it will prevail over the implicit system at every stage before the final reward is reached. To do this, it will need an accurate model of the strength of the implicit system's activation in any given modelled situation, as well as an accurate prediction of the outcome of the conflict between the implicit and explicit system. This amounts to assessing what the system as a whole *would* do in a given situation. An example might be a recovering alcoholic's plan for buying cigarettes during a downpour. Rather than walking two hundred yards in the pouring rain to a tobacconist's, he might decide to get some from the bar next door. As a plan, it is good as far as it goes — except that the proximity of immediate sources of alcoholic reward in the bar may cause his implicit system to suppress the planned action of leaving immediately in favour of that of buying a drink, and he should consider this before committing to the plan.

An Emergent Architecture: The Agent-Model and the World-Model

Looking at the above material, it is clear that the usefulness of modelling does not depend only on the formation and exploitation of models of the external world, but also on the modelling of many aspects of the agent itself. The agent must model the characteristics of its body and its sensors, and its ability to control the body as a function of sensory inputs, and to operate on objects and move through the environment. It may have to model the reward and punishment expected from modelled actions and situations; it may also have to model the behaviour of its own decision systems. As engineers, it is our working hypothesis that the best way of organising such a system will be for all the models of the various aspects of the agent to be linked functionally into a single composite model, the internal agent-model, which will have a degree of separation from all the models of the external world forming the world-model (Figure 6). Planning will then involve an ongoing interaction between the agent-model and the world-model, with each affecting the other; the actions of the modelled body will affect the modelled objects in the world-model, and the changes in the modelled objects as a result of those actions will cause corresponding changes in the modelled perceptions of the agent-model. Information about planned sequences of action with high evaluations must somehow be passed to the part of the system responsible for initiating actions, and these sequences must be selected sufficiently often and in appropriate circumstances to produce net benefits from the planning system, and to outperform competing designs.

The agent-model can directly affect only itself and the world-model; it can be directly affected by the world model, and by the agent qualities it models — the body, the sensors, the evaluations of the consequences of actions, and so on. In order to function well, it must be promptly updated with current information about the agent, so that it can accurately model and predict the agent's characteristics and behaviour. If it is a good model, and if the system for translating preferred plans into action is accurate, then the agent-model's preferred plans will be executed — presumably after a delay. The predicted changes to the body will occur, and the agent-model will be updated as predicted; the predicted changes in the world will occur, the world-model will be updated, and the updated world-model will affect the agent-model. The agent-model's plan will have been executed, and all the information coming in to the agent-model will be as predicted. In other words, the agent-model's actions in its original trial interaction with the world-model will produce the predicted effects in the updated agent-model and world-model. This virtual agent, trapped in its virtual world, will therefore appear to be able to act in the real world without in fact being able to do so.

If this separation between world-model and agent-model turns out to be advantageous or necessary in robots, then it invites the speculation that a similar separation may have happened in animals and humans. In other words, the efficient exploitation by evolution of the modelling abilities of the vertebrate brain may have led to an agent-model and a world-model as described above. If this has happened, then we suspect that the human agent-model will turn out to be the

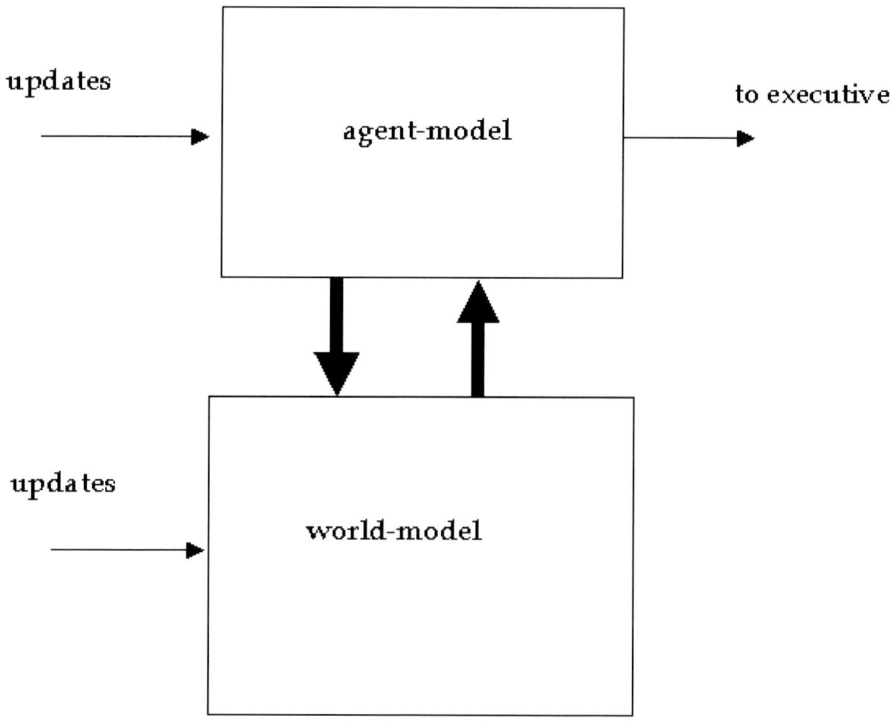

Figure 6. The hypothesised separation of the models involved in planning into two functional components, the agent-model and the world-model.

The agent-model operates on the world-model, and is itself affected by changes in the world-model. It may also be updated by other inputs, particularly those involved in the evaluation of the consequences of action. The world-model can be manipulated by the agent-model, and is also updated by sensory inputs. Information about favoured sequences of actions is passed from the agent-model to the executive, where it may influence the actions actually selected.

structure supporting conscious experience, and so the characteristics of conscious experience will be determined by the characteristics of the agent-model, and experience of the world may correspond to the agent-model's interactions with the world-model. We call this the IAM (Internal Agent-Model) theory of consciousness. Although the agent-model could equally well have been called the self-model, we wished to avoid the associations created by the many uses of the word 'self' in the context of consciousness. For example, Strawson (1999) distinguishes seven distinct aspects of the sense of self 'in so far as the self is experienced specifically as an inner mental presence' (Strawson, 1999, pp. 490–1); Damasio (1999) identifies the proto-self, the neural self, the core self, and the autobiographical self; and so on. By using the concept of the agent-model, we avoid identifying it with any particular aspect of the conscious self.

Interestingly enough, the philosopher Thomas Metzinger has proposed a theory of consciousness explicitly based around the concept of the self-model. Although his is primarily a phenomenological analysis, it is sometimes intriguingly close to the position we have taken:

> The phenomenal self is a virtual agent perceiving virtual objects in a virtual world . . . I think that 'virtual reality' is the best technological metaphor which is currently available as a source for generating new theoretical intuitions . . . heuristically the most interesting concept may be that of 'full immersion' (Metzinger, 2000).

He also notes that the phenomenal self-model

> is a plastic multimodal structure that is plausibly based on an innate and 'hardwired' model of the spatial properties of the system (e.g. a 'long-term body image'. . .) while being functionally rooted in elementary bioregulatory processes. . . .

The biologist Holk Cruse has also made an interesting contribution in this area (Cruse, 1999) and has ended up at a position very close to Metzinger's and to our own. His concern is with the possible structural and functional underpinnings of what he calls HIP systems (HIP = Having Internal Perspective), which he contrasts with NIP systems (Not having Internal Perspective). He begins by noting that the internal connections of simple feedforward neural network control systems 'can be interpreted as comprising an implicit world model (of that part of the world that is of importance for the system), because in some more or less indirect way it represents the properties of the world and the appropriate reactions' (Cruse, 1999). He notes that the inclusion of some recurrent (feedback) connections in such networks enables them effectively to predict sensory inputs, and to prepare the appropriate actions. He refers to both types of network (feedforward and feedback) as containing *non-manipulable* world models, in contrast to a third type of feedback network which can be isolated from its sensory inputs and motor outputs, and which can therefore 'play around in its virtual internal world' using what he calls its *manipulable* world model. Arguing that a world model that includes only 'information concerning the outer world' provides little support for a first-person internal perspective, he then proposes that things might be rather different in the case where 'some of the physical properties of the system itself are embedded into the internal world model' (Cruse, 1999). Drawing on some of his own previous work in neural networks for the control of movement (Steinkühler *et al.*, 1995) he sketches out a design for a system which contains a manipulable model of its own body. Finally, he advances his hypothesis:

> a system comprises a HIP system, i.e., has a first-person perspective, when (a) it contains a manipulable world model that includes properties of the system's own body which (b) can be used to compare the ('virtual') data of this model with those provided by the 'real' data from the sensory input . . . (Cruse, 1999).

Although his primary concern is clearly with the problem of natural consciousness, he does note that, if the hypothesis is correct, 'artificial systems could be constructed such as to have first-person perspective, i.e., to become HIP systems' (Cruse, 1999).

A number of recent papers and books have proposed the involvement in consciousness of some low-level neural structures representing and bringing together the physical, metabolic, and emotional state of the organism. In his review of Damasio's *The Feeling of What Happens* (Damasio, 1999), Douglas Watt summarised Damasio's position:

> consciousness requires that the brain must represent not just the object, not just a basic self structure, *but the interaction of the two.*

He went on to note:

> This is still an atypical foundation for a theory of consciousness, given that until recently, it was implicitly assumed that the self could be left out of the equation. There has been a recent sea change on this crucial point . . . (Watt, 2000).

A collection of papers on the self taken from several issues of the *Journal of Consciousness Studies* (Gallagher and Shear, 1999) gives a wide-ranging set of views on the nature of the self. Most are centred around the characteristics of the self as revealed through phenomenological analysis, but some regard the self as a structure or process, underpinned by neural systems, that provides some functional benefit to the organism as well as underpinning consciousness. Panksepp (1998) is an example:

> I advocate the position that the roots of the self go back to specific . . . sensory-motor action circuits within the mammalian brain which can generate a primitive sort of intentionality . . . and primitive forms of psychic coherence . . . by interacting with various emotional and attentional circuits that encode basic biological values. . . . These interacting circuits have specific neurochemical codes that may generate distinct types of neurodynamics within primitive core systems of self-representation that first symbolised organisms as coherently active creatures in the world (Panksepp, 1998).

Our proposal differs from the schemes of Panksepp, Metzinger, and Damasio in that the agent-model does not grow out of some primitive protoself or neural structure, but is instead a technical requirement of the very high level task of planning. Although their formulations are seductive, and elements of them could be accommodated within our scheme, we believe it will be best to pursue our approach in isolation so that it will be clear which properties of consciousness, if any, may derive exclusively from the planning requirement.

The Signs of Consciousness

Even if we succeeded in producing a robot system which successfully used an agent-model and a world-model for planning, how could we set about analysing the system in terms of possible links with consciousness? If we could observe the changes in the 'contents' of processes within the agent-model, and the relationship between those contents, the real world, and the actions of the robot, then we would be able to attempt to compare those characteristics of the agent-model with the characteristics of human consciousness: Baars' Global Workspace Theory might be a suitable framework to use (Baars, 1988). If they turned out — in the best possible case — to be practically identical, what would we be able to say? It would not be possible to claim that we had produced machine consciousness in the sense of Block's P-consciousness — we see no prospect of that claim ever being verifiable. It might be reasonable to claim, however, that we had identified the functional origins and components of the architecture within which consciousness exists, and that we had built a system with such an architecture. We would certainly be happy with this.

When consciousness is operating normally, there appears to be a close correspondence between subjective experience and reality. However, as is clear from examining the session headings in any modern consciousness conference, we now know that much of this correspondence is illusory. As Nørretranders puts it:

> Consciousness is a peculiar phenomenon. It is riddled with deceit and self-deception; there can be consciousness of something we were sure had been erased by an anaesthetic; the conscious I is happy to lie up hill and down dale to achieve a rational explanation for what the body is up to; sensory perception is the result of a devious relocation of sensory input in time; when the consciousness thinks it determines to act, the brain is already working on it; there appears to be more than one version of consciousness present in the brain; our conscious awareness contains almost no information but is perceived as if it were vastly rich in information. Consciousness is peculiar (Nørretranders, 1998, p. 286).

If the characteristics of the information flows through the agent-model in our system turned out to be peculiar in the same ways without our having deliberately made them so, then it would strengthen the claim that consciousness was the outcome of the operation of a similar system, and that the robot system might therefore represent a form of machine consciousness. Unfortunately, we do not believe that it is realistic to expect such high levels of performance from a robotic system in the near future, and so we must begin by looking for signs of more basic and non-illusory attributes of consciousness as set out by Baars.

Of course, engineers, like doctors, are familiar with the idea that a failing system can often reveal more about its principles of operation than can a system working normally. If you asked ten competent engineers to build a device conforming to some specification, you would be likely to end up with ten different devices. As long as they were all working normally under normal conditions, it would be difficult to tell the completed devices apart, or to make deductions about the design of each, because of course they would all behave identically. However, the different devices would be likely to fail in different ways, whether through structural defects or through operation outside normal conditions, and the ways in which they failed would give clues to their design — whether they used analogue or digital technology, whether they used closed loop or open loop control, and so on. If in the future we might wish to claim that our system was similar to that underpinning human consciousness, then it would be prudent to explore not only the similarities when both systems were working correctly, but also the similarities in failure modes when both systems were operating out of specification.

A Key Technical Challenge: Transparency

We have set out our reasons for supposing that models may be useful in achieving intelligence, and that some aspects of operation of certain models may have characteristics resembling those of consciousness. However, we need to find some way of making our systems transparent: unless we can devise some methods of showing that models have been formed, that they are models of certain other entities or processes, and that they are operating in certain ways, we will be unable to give an account of the system's internal workings. In the first experiments described above, it was possible to extract and represent the information in the models in ways that were useful in explaining the system's operation (Figure 2). Maintaining at least this degree of insight into internal processes as our systems grow in complexity is a key technical challenge of our programme.

Is there any general technique that can be employed to detect that a model is present, to show what the model corresponds to, and to show how it is being used? We are not aware of any such general approach, but two recent projects show that it may be possible to obtain such information on a case-by-case basis. In the first, Aleksander and his collaborators present a system including components closely modelled on the visual brain of humans in which the learned internal representation of each visual stimulus in the input array is constrained to form in such a way that it can be displayed in the same terms as the original stimulus - for example, the representation of a red square appears as a red square. (Aleksander *et al.*, 1999; 2001). After training, such internal representations can be elicited even in the absence of visual stimulation. When the system is presented with an ambiguous situation in which it has to produce a representation as a response, it is possible to see the progress and resolution of the conflict between alternative representations in terms of the visual stimuli from which they were originally formed. In the second, Edelman and his collaborators use the Darwin VII mobile robot to show how the process of neuronal group selection creates essentially arbitrary groups of cells whose coactivation indicates a response to a particular characteristic of visual input (for example, stripes) regardless of position, size, or orientation (Krichmar and Edelman, 2002). It is possible to see these groups being formed through experience, to see them being activated in the presence of the appropriate stimuli, and to appreciate their role in the control of the robot's behaviour. Although these projects form and use the equivalent of categorical representations, a relatively simple kind of model, they show the identification and characterisation of an internal representation used to inform or control decisions; we aim to extend this approach to deal with more dynamic models, and more complex manipulations and exploitations of such models.

One possible approach might be to take advantage of the fact that it is possible to have a complete record of all inputs, outputs, and internal processes for an artificial system, especially if it is a digital system. It might therefore be possible to train a neural network to identify, and perhaps even to characterise, the external and internal events that correspond to the states and state transitions of the various internal models. Presented with a sequence of activations from such an internal model, the neural network would be required to yield the events in the world that the model was representing. There are many reasons why this might prove to be impossible, but the potential rewards are so great that we feel it should at least be attempted.

Acknowledgments

The ARAVQ simulations and experiments were carried out at the California Institute of Technology by Lerone Banks and Nathan Grey.

References

Aleksander, I., Dunmall, B. and Del Frate, V. (1999), 'Neurocomputational models of visualisation: A preliminary report', *IWANN (1), 1999*, pp. 798–805.
Aleksander, I., Morton, H. and Dunmall, B. (2001), 'Seeing is believing: Depictive neuromodeling of visual awareness', *IWANN (1), 2001*, pp. 765–71.
Arbib, M.A. (1972), *The Metaphorical Brain: An Introduction to Cybernetics as Artificial Intelligence and Brain Theory* (Chichester: Wiley-Interscience).
Baars, B.J. (1988), *A Cognitive Theory of Consciousness* (Cambridge: Cambridge University Press).
Behrmann, M. (2000), 'The mind's eye mapped onto the brain's matter', *Trends in Psychological Science*, **9** (2), pp. 50–4.
Block, N. (1995), 'On a confusion about a function of consciousness', *Behavioral and Brain Sciences*, **18** (2), pp. 227–87.
Brooks, R.A. (1991), 'Intelligence without representation', *Artificial Intelligence*, **47**, pp. 139–59.
Churchland, P.S. (2002), 'Self-representation in nervous systems', *Science*, **296**, pp. 308–10.
Clark, A. and Grush, R. (1999), 'Towards a cognitive robotics', *Adaptive Behavior*, **7** (1), pp. 5–16.
Craik, K.J.W. (1943), *The Nature of Explanation* (Cambridge: Cambridge University Press).
Crick, Francis (1994), *The Astonishing Hypothesis: The Scientific Search for the Soul* (New York: Charles Scribner's Sons).
Cruse, H. (1999), 'Feeling our body: The basis of cognition?', *Evolution and Cognition*, **162** (5), p. 2.
Cyberbotics (2003a), *http://www.cyberbotics.com* Web pages for Webots; Cyberbotics Sarl, Lausanne.
Cyberbotics (2003b), *http://www.cyberbotics.com/products/robots/khepera.html* Web pages for the Khepera; Cyberbotics Sarl, Lausanne.
Damasio, A.R. (1999), *The Feeling of What Happens: Body and Emotion in the Making of Consciousness* (London: Harcourt Brace & Co).
Dawkins, R. (1976), *The Selfish Gene* (Oxford: Oxford University Press).
Dawkins, R. (1998), *Unweaving the Rainbow* (London: Penguin).
Dennett, D.C. (1995), *Darwin's Dangerous idea: Evolution and the Meanings of Life* (London: Allen Lane, The Penguin Press).
Dreyfus, H.L. (1992), *What Computers Still Can't Do* (Cambridge, MA: MIT Press).
Edelman, G.M. (1987), *Neural Darwinism: The Theory of Neuronal Group Selection* (New York: Basic Books Inc.).
Gallagher, S. and Shear, J. (ed. 1999), *Models of the Self* (Exeter: Imprint Academic).
Grush, R. (1997), 'The architecture of representation', *Philosophical Psychology*, 10, pp. 5–23.
Grush, R. (2002), 'An introduction to the main principles of emulation: Motor control, imagery, and perception', Technical Report, Philosophy, UC San Diego.
Hauser, M.D. (2000), *Wild Minds: What Animals Really Think* (New York: Henry Holt).
Hesslow, G. (2002), 'Conscious thought as simulation of behaviour and perception', *Trends in Cognitive Sciences*, **6**, pp. 242–7.
Johnson-Laird, P.N. (1983), *Mental Models: Towards a Cognitive Science of Language, Inference, and Consciousness* (Cambridge: Cambridge University Press; Cambridge, MA: Harvard University Press).
Kelly, I., Holland, O., Scull, M. and McFarland, D. (1999), 'Artificial autonomy in the natural world: Building a robot predator', *Proceedings 5th European Conference on Artificial Life (ECAL'99)*, pp 289–93 (Berlin: Springer-Verlag).
Krichmar, J.L. and Edelman, G.M. (2002), 'Machine psychology: Autonomous behavior, perceptual categorisation, and conditioning in a brain-based device', *Cerebral Cortex*, **12**, pp. 818–30.
Linåker, F. and Niklasson, L. (2000a), 'Extraction and inversion of abstract sensory flow representations', in *Proceedings of the Sixth International Conference on Simulation of Adaptive Behavior: From Animals to Animats 6 (SAB2000)*, pp. 199–208 (Cambridge, MA: MIT Press).

Linåker, F. and Niklasson, L. (2000b), 'Time series segmentation using an adaptive resource allocating vector quantization network based on change detection', in *Proceedings of the IEEE-INNS-ENNS International Joint Conference on Neural Networks (IJCNN 2000)*, pp. 323–8 (IEEE Computer Society).
Metzinger, T. (2000), 'The subjectivity of subjective experience: A representationalist analysis of the first person perspective', in *The Neural Correlates of Consciousness*, ed. T. Metzinger (Cambridge, MA: MIT Press).
Minsky, M.L. (1968), 'Matter, mind, and models', in *Semantic Information Processing*, ed. M.L. Minsky (Cambridge, MA: MIT Press).
Nauck, D., Klawonn, F. and Kruse, R. (1997), *Foundations of Neuro-Fuzzy Systems* (Chichester: Wiley).
Nørretranders, T. (1998), *The User Illusion: Cutting Consciousness Down To Size*, trans. J. Sydenham (London: Allen Lane, Penguin Press).
Panksepp, J. (1998), 'The periconscious substrates of consciousness: Affective states and the evolutionary origins of the self', *Journal of Consciousness Studies*, **5** (5–6), pp. 566–82.
Passino, K.M. and Yurkovich, S. (1998), *Fuzzy Control* (New York: Addison-Wesley).
Perlis, D. (1997), 'Consciousness as self-function', *Journal of Consciousness Studies*, **4** (5–6), pp. 509–25.
Ramachandran, V.S. and Blakeslee, S. (1998), *Phantoms in the Brain: Human Nature and the Architecture of the Mind* (London: Fourth Estate).
Richter, W., Somorjai, R., Summers, R., Jarmasz, M., Menon, R.S., Gati, J.S., Georgopoulos, A.P., Tegeler, C., Ugurbil, K. and Kim, S.G. (2000), 'Motor area activity during mental rotation studied by time-resolved single trial fMRI', *Journal of Cognitive Neuroscience*, **12** (2), pp. 310–20.
Rolls, E.T. (1999), *The Brain and Emotion* (Oxford: Oxford University Press).
Sporns, O., Almássy, N. and Edelman, G.M. (2000), 'Plasticity in value systems and its role in adaptive behavior', *Adaptive Behavior*, **8** (2), pp. 129–48.
Steinkühler, U., Beyn, W-J. and Cruse, H. (1995), 'A simplified MMC model for the control of an arm with redundant degrees of freedom', *Neural Processing Letters*, **2**, pp. 11–15.
Strawson, G. (1999), 'The self and the SESMET', in Gallagher and Shear (1999).
Tani, J. (1998), 'An interpretation of the "self" from the dynamical systems perspective: A constructivist approach', *Journal of Consciousness Studies*, **5** (5–6), pp. 516–42.
von Uexkull, J. (1934/1957), 'A stroll through the worlds of animals and men', in *Instinctive Behaviour: The Development of a Modern Concept*, ed. C.H. Schiller and K.S. Lashley (New York: International University Press).
Watt, D.F. (2000), 'Emotion and consciousness II: A review of "The Feeling of What Happens"', *Journal of Consciousness Studies*, **7** (3), pp. 72–84.
Werbos, P.J. (1990), 'A menu of designs for reinforcement learning over time', in *Neural Networks for Control*, ed. R.S. Sutton, W. T Miller, III and P.J. Werbos (Cambridge, MA: MIT Press).
Werbos, P.J. (1992), 'Approximate dynamic programming for real-time control and neural modeling', in *Handbook of Intelligent Control*, ed. D. White, A. and D.A. Sofge (Van Nostrand Reinhold).
Wilkinson, S. (2000), 'Gastrobots: Benefits and challenges of microbial fuel cells in food powered robot applications', *J. Autonomous Robots*, Sept. 2000
Wolpert, D.M. and Ghahramani, Z. (2000), 'Computational principles of movement neuroscience', *Nature Neuroscience Supp. vol 3*, pp. 1212–17.

Jesse J. Prinz

Level-Headed Mysterianism and Artificial Experience

I: Introduction

Many materialists believe that we should, in principle, be able to build a conscious computing machine. Others disagree. I favour a sceptical position, but of another variety. The problem isn't that it would be impossible to create a conscious computer. The problem is that we cannot know whether it is possible. There are principled reasons for thinking that we wouldn't ever be able to confirm that allegedly conscious computers were conscious. The proper stance on computational consciousness is agnosticism. Despite this agnosticism, I think we are very close to understanding the material basis of consciousness. Close, but we will never get all the way there. Our understanding of the material basis of consciousness is ineluctably incomplete. That makes me a mysterian. But I am not a defeatist mysterian. I do not think the irresolvable mysteries of consciousness prevent us from formulating concrete empirically grounded theories of consciousness. I will even outline such a theory below. I also think we can identify some properties that are necessary for consciousness, and some that are sufficient. The problem is we cannot find properties that are both. This may sound like a contradiction, but it is not. We can know necessary properties and sufficient properties without knowing necessary and sufficient properties. One can know that drinking two litres of alcohol is sufficient for intoxication, and that at least one teaspoon of alcohol is necessary without knowing the quantity that is both necessary and sufficient. In the case of consciousness, the properties we know to be sufficient are properties that computers lack. The properties we know to be necessary can be possessed by computers, but we don't know whether those properties suffice. We can equip computers with every knowable necessary property and still scratch our heads when asked whether we have managed to create artificial experience.

Before embarking, I want to get slightly more precise about the question. We need a working definition of what consciousness is and what machines are. By consciousness, I shall be referring to qualitative, phenomenal feelings — Nagel's famous what-it's-likes. I shall talk of conscious states and experiences interchangeably. If there is a notion of consciousness that involves nothing but information access, as Block (1995) has argued, then it is highly likely that computers can be conscious. My laptop is conscious of what letters I type into it, and it dutifully reports these to me on its screen. Against this possibility I would deny that mere access, without phenomenal qualities, qualifies as consciousness. That said, I actually think access *contributes* to phenomenal qualities. The question that I am addressing asks whether the forms of access available to computers can be sufficient for phenomenology.

Computers are automatic formal symbol manipulators (Haugeland, 1985). This definition is liberal enough to include pocket calculators and artificial neural nets. The latter may not appear symbolic at first, but I take a symbol to be an object that is in the business of representing something, and a formal symbol system to be a system in which symbols have causal effects in virtue of their formal (or physical) properties. Most connectionists should agree that nets are computers in this sense. We, too, are computers at some level of description. Patterns of neuronal activation are symbolic. Thus, it follows trivially that at least one kind of computer can be conscious: the human brain.

To escape triviality, I would follow McGinn (1999) in defining the question of computer consciousness as the question of whether we can create consciousness out of inorganic stuff. This is roughly equivalent to the question: could our brains be conscious in virtue of their computational properties rather than their structural/material properties? This is what I have in mind. I will critically assess two standard strategies for delivering a sceptical response, both of which have kernels of truth. I will then sketch a scientifically plausible theory of consciousness to see whether it helps us determine whether inorganic brains can be conscious. That exercise will serve to demonstrate that we can make great progress in our understanding of consciousness while still making little progress on the question of conscious machines. We must be mysterians, I will argue, but of a more level-headed variety than typically described.

II: Two Standard Arguments Against Machine Consciousness

There are various ways in which a materialist might resist the conclusion that we can construct conscious computers. Two lines of argument are especially alluring. According to the first, computers don't have what it takes. Consciousness relies on properties that computers simply lack. In particular, computers lack biological properties, properties that are found in the nervous systems of carbon-based life forms. According to the second line or argument, there are epistemic barriers to building conscious computers. We just can't figure out how to do it. More specifically, we cannot figure out how to build a conscious computer because we cannot figure out how consciousness works in us. Let's consider these theses in turn.

The biology thesis

Some consciousness researchers are committed to a biologically based form of materialism. They assign a special role to the physical materials found in mammalian brains. Searle (1980; 1992) defends something close to this view. Strictly speaking, Searle says that biology *could* be essential to consciousness without saying that it actually is necessary. So he isn't a perfect representative of the view I have in mind. But Searle does say that computational properties of brains cannot be sufficient for consciousness. What's the missing ingredient? It *could be* biology. Here's a version of an argument that can be reconstructed from Searle (1992). Consciousness is an intrinsic property of conscious minds; it does not depend on the interpretation of others. Computational properties are not intrinsic, because syntax and semantics cannot be defined in terms of physics. A brain counts as a computer only in virtue of being interpreted in a particular way. Biological properties, in contrast, are intrinsic. Therefore, consciousness could not be a computational property, but it could be a biological one. Strictly, Searle's view is that consciousness *emerges* from biological properties; it is a *causal* effect of brain events, not to be identified with brain events.

We also see traces of the biology thesis in Block's (1978a) absent qualia arguments. Block has us imagine entities whose functional properties match those found in the human mind, but realize those functional properties via very different kinds of stuff. The Bolivian economy, Block conjectures, could accidentally conform to a Turing table description of the input/output functions computed by a human mind. It is an affront to intuition, he continues, to conclude that the economy of Bolivia would thereby have qualitative states. Block's views differ in important ways from Searle's. For example, he allows that computation may be sufficient for *intentionality*, and he seems open to the possibility that phenomenal properties can be identified with, rather than merely caused by, biological properties. Both authors are receptive to the view that consciousness may depend on biology.

I cannot present a full assessment of Searle and Block here, but I offer a few remarks to make it clear where I stand. I do not accept Searle's premise that computational properties are interpretation-dependent. Any system that has states that stand for something, and processes that govern those states in virtue of their form, is a computer. Some systems have symbols that stand for something in virtue of being interpreted that way. But this is not an essential property of a computer. Consider a system that has internal states that reliably detect a certain property P and only P, and it uses these states to orchestrate behavioural interactions with things that instantiate P. I would call such a system a computer, and I would deny that its status as such depends on us. There are two kinds of cases where I think we should regard properties as interpretation-dependent. There are some cases where a property is explicitly defined as dependent on us. Being scary and being a suspect are examples. There are other cases where our interpretations are the only source of commonality or coherence in a category. Some people think that colours are interpretation-dependent in this sense, and words in public languages may also qualify. But, being a computer is not interpretation-dependent in either sense. Computation is not defined with reference to human responses, and when we assign computational interpretations to things, there are systematic properties intrinsic to those things that make the computational interpretations successful. As far as I can tell, Searle bases his conclusion that computation is interpretation-dependent on the premise that syntax and semantics cannot be defined in terms of physics. That strikes me as quite a leap. I think computational properties exist without interpretation and, as such, they are candidates for being the bases of phenomenal qualities.

Now consider the absent qualia argument. Intuitions strongly support Block's contention that the economy of Bolivia lacks conscious states. The difficulty is that we cannot let intuitions arbitrate in such matters. Once we assume that consciousness has a material basis, we must be open to the possibility that it can appear in unexpected places. The intuition that the economy of Bolivia is *the wrong kind of thing* to have conscious states may owe to residual dualism. We are forced, in some sense, to say brains are conscious, but we want to resist ascribing consciousness to physical systems when we are not forced to. That said, I do not think the economy of Bolivia has conscious states. But the reason for my doubts, and I suspect Block would agree, have more to do with what I understand about the basis of consciousness in us. Conscious states are amazingly rich. Very minute differences in perceived objects can lead to differences in conscious experience. These differences can be captured at the level of changes in small populations of neurons. Conscious states also have very specific temporal dynamics and causal roles. These facts may depend on very specific principles of organization found in our nervous systems. The contents and temporal dynamics of some conscious states can be mapped onto specific regions of the brain. For example, motion illusions are well correlated with activity in medial temporal cortex (Tootell *et al.*, 1995), and experiences of colours and shapes are correlated with cellular responses in ventral extrastriate areas (Zeki, 1993). In order to have states that were functionally isomorphic with our conscious states (isomorphic in their temporal, causal and content properties), a system might have to be as complex as our brains, at the level of small populations of neurons. Since it is unlikely that the economy of Bolivia is isomorphic with events at a level that we know to have an impact on consciousness in us, we lose reason for thinking that it is conscious.

These considerations leave open the possibility that the economy of Bolivia might be conscious if it were isomorphic to the neural dynamics in the human brain. Perhaps biological *materials* are not necessary for consciousness. The things that seem to impact consciousness in us are dynamical interactions within and between populations of neurons, and we have no guarantee that the chemical constitution, electrical profile or morphology of those neurons is an essential part of the picture. A population of artificial neurons in a suitably programmed computer might be isomorphic with the level that seems to matter. A full brain simulation in a vast artificial neural network would have artificial cells that could be activated, deactivated and causally connected in ways that might allow a range of states and temporal dynamics isomorphic to the conscious states that we report. For every reported change in our conscious states, there might be a change in such a system, and these changes might arise under the same stimulus conditions. I will return to this issue below.

The interim moral is that it might be wrong to claim that consciousness requires some property describable only in the language of biology. But, for all I have said, it might be right. Just as we cannot conclude that biological materials are essential for consciousness, I have not ruled out the possibility that they are.

The ignorance thesis

Turn now to the second kind of argument against computer consciousness. Sceptics of this kind argue that we cannot create artificial experience because we do not know enough about how experience comes about in our own case. McGinn (1991; 1999) seems to hold a view of this kind. He proposes that human beings suffer from a basic epistemic limitation. Just as dogs cannot understand calculus, we can never understand how feeling is related to matter. We can examine the brain thoroughly, in all its functional and structural detail, and still wonder how consciousness arises. That puzzlement may be explained by the supposition that the problem of consciousness has a solution beyond the borders of our epistemic dominion. McGinn does not reject materialism. He says that consciousness is probably a consequence of some physical property, P. In fact, he thinks P is a biological property, which puts him in the same camp as Searle and possibly Block. But McGinn insists that we do not and cannot know what P is. If P is unknown to us, we cannot design a computer to have P.

It is not very satisfying to simply assert that the problem of consciousness is insoluble. We should have some principled reason for thinking it is insoluble. After all, it is a problem we can formulate. Dogs are essentially incapable of using external symbols, much less mathematical equations. Their problems with calculus are a consequence of a much more general limitation in how they think. Of course, most problems are easier to formulate than to solve. That's what keeps scientists and philosophers in business. Often, new discoveries and concepts are needed to make progress on difficult problems. Unlike our non-human cousins, we seem to be especially adept at doing that. Arduous intellectual toil furnishes us with the resources to explain some of the hardest puzzles of the universe. Many of the efforts are works in progress, and some have involved the deployment of concepts that are a difficult to grasp. A little exposure to modern cosmology can drive home that point. But one doesn't have the sense that cosmology will never admit of satisfactory theories. Physicists have already developed theories that seem to give some explanatory purchase on the basic building blocks of being, at least in some mathematical and metaphorical sense. Given all this, one wants to know what intellectual deficit keeps consciousness out of reach. Is it that math and metaphor are bad tools for building epistemic bridges from brain mush to qualia? The real problem, says McGinn, is that we just cannot see the brain for what it is. Some of its properties are invisible to us. If we could understand the brain fully, if we could see it from God's point of view, we would immediately see how it gives rise to consciousness. For McGinn, it's not that we have discovered the property that links mind and brain, but haven't identified it as such. It's not like the police inspector who has found the right suspect but can never find proof of guilt. McGinn says the right suspect eludes us.

I am sympathetic to McGinn's sceptical conclusion, but I part company with him on several other points. First, I think we can understand the brain quite fully. The mind–body problem does not stem from an inability to see the brain in all its glory. There is no hidden property that escapes detection when we turn on our magnets, implant electrodes or stain chunks of brain. There may be properties we haven't managed to observe yet, but there are no intractably unobservable properties (except those pesky quantum phenomena that are a refuge for non-mysterians with an appetite for the mysterious).

Second, McGinn thinks that certain theories of the physical basis of consciousness can be ruled out. In particular, he sides with Searle in claiming that consciousness is not a computational property. I am less confident. I think that consciousness could be a computational process. We do not have good grounds to rule this out.

This brings me to a third disagreement with McGinn. With many others, McGinn thinks that the mind–body problem would have to be solved by providing an *explanation*. Some physical property would have to explain qualitative feels. I deny this. Materialists need to find an identity. They need to find the physical (or computational/functional) property that *just is* qualitative feels. Identities are not the kinds of things we can explain (see also Block, 1978b; Papineau, 2002). We don't ask, Why is Charles Dodgson identical to Lewis Carroll? Instead, we ask an epistemic question. How can we be confident that Dodgson is Carroll? It is here that explanation comes in. Identities gain support when they explain certain things. One might say I know Dodgson is Carroll because he keeps getting royalty checks from the publisher of *Through the Looking Glass*, or they both report to be out of the office at the same times. To support a mind–brain identity claim, we want to show that there are correlations between the two, and that the correlations give us some explanatory buying power. Or, more accurately, we want to show that the correlations would lead to an elegant explanation of certain phenomena if we assume that the correlated things are identical. Identity claims that are defended in this way *serve* as explanations. They are not things to be explained in their own right.

My fourth departing stride from McGinn follows closely on the heels of the third. I think that we can actually make considerable progress on the problem of consciousness. Correlations and explanatory identities lead us to good, concrete hypotheses about the material basis of consciousness.

This last declaration looks like a repudiation of mysterianism, but it isn't intended to be. I think there is a deep sense in which the mind–body problem is insoluble, and this compels me to be agnostic about artificial experience. We can learn a lot about the material basis of consciousness, but not everything. Like McGinn, I think that artificial experience eludes us because the basis of experience eludes us. But the basis of experience eludes us for humbler and less humbling reasons than McGinn would have us believe. The epistemic ceiling has little to do with human cognitive limitations. Before getting to this epistemic ceiling,

I'd like to consider the walls of wisdom. It is instructive to see just how much we know about the material basis of consciousness before hitting the point where knowledge stops.

III: What We Know

A theory of consciousness

Those who want to probe the possibility of conscious machines should begin by looking to our best theories of how consciousness arises in human beings. Investigation into this question might promote a degree of optimism. We are learning quite a bit about the material basis of consciousness. It is only a matter of time before we find a theory that gains widespread support within scientific communities. Disputes will inevitably continue, but we are probably moving toward a point where the most active disagreements concern details rather than the big picture. If we arrive at such a point, the question of machine consciousness might appear quite tractable. We can take the theory of human consciousness and ask whether the properties it describes can be implemented in machines. We are still far from the end of inquiry, but we are in a position to see how this sort of investigation might go. I want to briefly present what I take to be the best current empirically informed theory of consciousness. I have offered more detailed discussions and defence elsewhere (Prinz, 2000; 2001). The theory will serve to illustrate a disappointing moral. It simply isn't the case that scientific investigations into the nature of consciousness will make questions of machine consciousness disappear. Even if scientific theories of consciousness succeed by their own standards, we must remain agnostic about artificial experience.

Theories of consciousness are sometimes stated at a single level of analysis. Philosophical theories are often expressed in folk psychological terms (see, for example, Rosenthal's (1986) higher-order thought theory, or other functionalist accounts). Other theories are pitched at lower levels. Consider, for example, the hypothesis that consciousness involves neuronal oscillations around 40 Hz (Crick and Koch, 1990). But it is difficult to present a theory at one level without having an eye to other levels. The oscillation theory is supposed to account for the higher-level fact that components of a conscious percept are bound into a unified whole. So the theory might be expressed by saying that consciousness arises when the parts of a percept cohere. Ascending from the brain to the level of psychology is important if we want to say what consciousness is for. Conversely, theories pitched at the level of psychology sometimes make reference to events at lower levels. Role-filler functionalism (associated with David Lewis) identifies conscious states with neural states that play particular psychological roles. Some think that conscious states could not be causally efficacious if they are roles, rather than fillers of roles. Rosenthal's higher-order thought theory must also push beyond the folk psychological level if placed under pressure. If thoughts are just propositionally structured representations, as folk psychology might have it, then simple computers can have higher-order thoughts. To avoid being too liberal about consciousness, defenders of these theories might want to place more constraints on what thoughts are and those constraints might be stated in the language of a scientific psychology or, perhaps, in neuroscientific terms.

This multi-levelism is important, for reasons that I will explore further below. Accordingly, I will present the theory of consciousness that I favour, at more than one level of analysis. It is sometimes pointed out that there are many levels of analysis to chose from (see Churchland, 1986), but it is useful to begin with three. I want to use something like Marr's (1982) scheme. At the top, we can identify a level that might be called the psychological profile (an improvement over Marr's confusing term 'computational'). The psychological profile uses the language of folk psychology to characterize three important dimensions of our mental states. It says what those states represent, how or when they come about, and what function they serve. Below the level of psychological profile, we have Marr's algorithmic level. It is the level at which a theoretician would postulate rules and representations, or mechanisms of information flow between postulated information-processing systems. It is the level at which most models in scientific psychology are situated. And, finally, we have a level (itself stacked up from many levels) of neuronal implementation. Here, the mechanisms at the algorithmic level are mapped onto structures and processes that can be described using the vocabulary of some branch of neuroscience.

We address questions about consciousness at any of these levels. Beginning with the first, we can ask, what are the contents of consciousness? Under what conditions does consciousness arise? And what roles does it play? I offer the following answers:

The contents of consciousness are sensory (specific to individual sense modalities) and they present the world from a particular vantage point. We cannot have a conscious experience of a view from nowhere.

Consciousness arises when we are paying attention. We seem to lose awareness of one thing when we intensely focus on another. This phenomenon has been experimentally confirmed by Mack and Rock (1998). They present subjects with the attention-demanding task of trying to discern which of two intersecting lines is longer. While subjects perform this task, an unexpected stimulus (such as a geometrical shape) is flashed in plain view. Many subjects fail to experience that stimulus — at least that's what they claim when asked. Mack and Rock conclude that attention is necessary for consciousness.

Consciousness is involved in flexible response. Responses that are overlearned or performed without variation can be executed without awareness. Consciousness makes information available for deliberation and deliberate control of action. Anecdotally, this is supported by the fact that our conscious states always seem to be available for reporting, even if we do not actually report them.

These three points hang together. In the theory of consciousness that I have defended elsewhere, I argue that consciousness arises when perspectival perceptual states become available for deliberate control as a result of attention. So far, this theory is stated at the level of psychological profile, but it can also be told at other levels.

At the algorithmic level, one might begin by identifying the rules and representations corresponding to perspectival perceptual representations. As Ray Jackendoff (1987) notes, computational models tend to assume that perceptual systems progress through stages of representation that are increasingly abstract. The lowest-level representations correspond to very local features without capturing how these hang together to form unified wholes. The highest level compensates for that deficiency by generating representations that abstract away from specific details of vantage point (such as scale, metric proportion, relative orientation, etc.). The intermediate level gives us representations that are vantage-point specific and coherent. Jackendoff says that consciousness must be located at this intermediate stage of information processing. Because it depends on models of perception that go beyond observable features, this story is pitched at the algorithmic level. One can also tell an algorithmic-level story about attention appealing to our best models of how stimulus selection works. Those models typically assume that attention is a mechanism by which representations are filtered for further processing. This brings in the third component of the theory I have presented. Attention allows information to flow to centres responsible for deliberate control. The prevailing view in psychology is that deliberate control is under the jurisdiction of working memory, a temporary storage faculty that has various executive abilities, which can be described in terms of rules and representations (Baddely and Hitch, 1977).

At the algorithmic level, consciousness arises when intermediate-stage representations in hierarchical perceptual systems send outputs, by surpassing some specifiable threshold for attention, to executive working memory systems. A full algorithmic description would flesh out each of these components using specific models. We can go a step further by unpacking those models in neural terms. Hierarchically organized perceptual systems have been correlated with anatomical regions. Vision, for example, includes a hierarchy that moves from primary visual cortex, into extrastriate areas, and then up to posterior parietal and inferotemporal cortices. Attentional mechanisms have been associated with dynamic changes in connectivity between neural populations achieved by changing receptive fields or modulating sensitivity through gain control (e.g., Moran and Desimone, 1985). In vision, attention may be driven in part by mechanisms in inferior parietal cortex. Executive working memory is generally thought to involve regions in frontal cortex (e.g., Haxby *et al.*, 2000). Consciousness, then, will typically involve a circuit that includes perceptual centres in temporal cortex, attentional centres in parietal cortex, and working memory centres in frontal cortex, which is exactly what functional imaging studies reveal (Rees, 2001). Each system in these circuits contains networks of neurons whose response profiles, chemical receptors and connectional anatomy can be described in considerable detail.

Following these lines, consciousness researchers are poised to give a highly detailed theory of the material basis of consciousness in human beings. On the face of it, this should fuel optimism with regard to machine consciousness. If we can identify the physical basis of consciousness in us, then it seems more likely that we will be able to build conscious machines. At the very least, a good theory of the material basis of human consciousness should render the question of whether conscious machines can be created much easier to answer. Such optimism is sadly misplaced.

IV: Levels and Ignorance

For purposes of discussion, I will assume that the foregoing theory of consciousness is true. It will help me to show that progress in the science of consciousness may offer little help to those who want to engineer consciousness. If I am right, consciousness involves a number of systems: perception, attention and working memory. This makes the task of modelling consciousness daunting, but complexity is not grounds for scepticism. The real difficulty stems from the fact that we can describe the key systems involved in consciousness at varying degrees of abstraction. Working memory might be described as a temporary storage faculty with some executive capacities, or it might be described as a temporary storage faculty of a particular type with a particular set of rules, representations, operations and so forth. These rules and representations are carried out by means of physical events in the nervous system. In short, working memory can be analyzed at the level of psychological profile, algorithm or implementation. The same is true of perception and attention. In each case, it is far from clear how we should determine which levels matter.

The problem of levels is compounded by the fact, brushed aside earlier, that there are in reality many more than three levels of analysis. Each of Marr's levels is itself multi-tiered. The psychological profile can be characterized in terms of coarse functions or specific mental faculties. The algorithmic level can be described using such tools as classical models, neural nets and differential equations. Such models can offer perfectly compatible descriptions of the same system. And the implementation level can be characterized in terms of large-scale neural networks, the spiking patterns of individual neurons, chemical transactions at synapses and so forth. There are scores of levels to choose from. This would not be a problem in and of itself, but it is exacerbated by two other considerations.

The individuation problem

First, there is no established method for individuating levels. Levels are, in part, a function of our own descriptive practices. They are a function of how much detail we want to apply when explaining a mechanism (see Craver, 2001). This does not mean that levels are not real. Counterfactual supporting generalizations exist at different levels of detail. Moreover, there is a straightforward sense in which we can talk about levels mereologically. Think of a mechanism as a physical device that produces outputs given inputs. Now think of that device as performing this transformation in a certain way. For example, it might take several steps to compute the input/output function, and each step might perform some important subtask. The steps are carried out by mechanisms; they depend on physical devices that implement input/output functions. These mechanisms might be said to reside at a lower level of analysis, and they might decompose into still lower mechanisms. But notice that there are no fixed levels that cut across all physical systems. Simpler devices may decompose into fewer mechanistic parts, and there is no principle in virtue of which different levels in different devices are the same. The number of levels is determined by the mechanistic complexity of a system, not by fixed strata in nature. One possible exception to this are cases in which two mechanisms are defined over inputs and outputs that are described using vocabulary in the same branch of science. But, even here, there is no guarantee of convergent levels. For example, chemical processes in etching may not belong to the same explanatory level in any interesting sense as the chemical processes that are involved in drawing or lithography. With regard to brains and computers, things are just as bad. There is no reason to expect perfectly convergent levels when comparing inorganic and organic devices. Chemistry may play an important role in explaining things at the level of inter-neuronal communication in brains, but it is not crucial to appeal to chemistry in order to explain how microcircuits communicate. Does this show that chips and cells reside at different levels of analysis? The answer is far from clear. If we cannot always find levels in computers that correspond to levels in human minds, then we might discover the level at which consciousness arises in minds without having any clear idea about whether consciousness can be created in machines.

The difference-maker problem

Can we even identify the level at which consciousness arises in humans? One might think the answer to this question can be ascertained experimentally. The trick would be to alter activity at one level, while keeping other levels constant. Any level at which an alteration affects consciousness is a level at which processes necessary to consciousness may reside. Call this the method of difference-makers. We look for the level(s) at which changes make a difference to consciousness.

One difficulty with this method is that it is be impossible to change a high-level process while keeping lower levels *completely* fixed, because, in the scheme I have described, high levels supervene on lower levels. One could, however, change the high level while keeping some aspect of lower levels fixed. So, for example, one can ask whether activity in neurons (implementation level) that normally realize perceptual representations (psychological profile level) would be associated with conscious states if they were active when one was not paying attention (psychological profile, again). The answer here seems to be 'no'. Cells in perceptual systems that are active during conscious experiences can fire when one is not paying attention. But, when this occurs, evidence suggests that one has no conscious experience associated with those cells. Mere activity in particular cells is not sufficient for consciousness. That activity comes into consciousness only if it realizes a particular kind of psychological state: attentive perception. Reasoning in this way suggests that the psychological profile level captures a level of processing that is essential for consciousness. This should come as no surprise. In characterizing the psychological profile level, we were implicitly using something like the method of difference-makers but applied at a single level. When we asked, 'under what conditions specified in the language of folk psychology does consciousness arise?' attention emerged as a plausible candidate. Attention seems, in anecdote and experiment, necessary for consciousness. If our best theory of attention has it that attention just is a gate control to executive faculties, then the necessity of attention for consciousness also entails the necessity of executive faculties as well.

One might resist these necessity claims by insisting that, for all we know, there are possible creatures that have phenomenology but lack attention and executive working memory. Such a hypothetical creature would have a radically different psychology but have experiences, albeit experiences that may differ from our own. I do not think there is any reason to take this possibility seriously. While we cannot rule out consciousness in creatures whose behaviour is driven by radically different kinds of mechanisms, we have no reason to think such creatures are conscious. There is no theoretical pressure to attribute consciousness to plants or single-celled organisms, for example. If one found creatures that exhibited the kind of flexibility afforded by human consciousness, it might become very tempting to postulate consciousness. But notice that, in order to have flexibility of the kind we find in humans or other advanced mammals, a creature would have to be driven by a cognitive architecture that had something like faculties for deliberation and attention. Thus, the psychological profile of such a creature would, in fact, be like ours. Without deliberation, a creature would be driven by immediate reflexes, and, without attention, it would not be able to selectively sample from perceptual inputs. Such a creature would have to have perceptual systems that did a lot of selection without attention, say, having receptors that are restricted to a very limited range of inputs. Paramecia, clams and insects may have 'psychologies' like this. But we find little compulsion to insist that such creatures are conscious.

Attention and executive faculties are, I conclude, plausible candidates for necessary conditions on consciousness. I would also add perception to this list. All human consciousness is consciousness of perceptual states (I would argue), and there is no reason to think one can be conscious of anything else. I say this, in part, because I think all mental representations are perceptual in origin, but that is a very controversial claim (Prinz, 2002). It seems quite obvious that changing perception, while keeping other things fixed, has an impact on awareness. If you attentively stare at a blue wall and then attentively stare at a red wall, you will have different experiences. So consciousness changes with perceptual states. Perception makes a constitutive difference to consciousness. This is not to say that a creature can be conscious only if it perceives what we do. Other creatures may be perceptually sensitive to things we cannot see. There are even individual perceptual differences within our own species. The method of difference-makers suggests that perceptual representations *of some kind or other* are essential to consciousness.

But what happens when we descend to lower levels of analysis? Suppose we talk, not of attention, deliberation and perception, but of the specific algorithms by which these faculties operate in us. Must perceptual systems be divided up the way ours are? Must they use representational primitives of roughly the same kind? Must attention work by the same selective principles? Must the same items pop out? Must the same number of objects be selectable? Must top-down templates be available for attention? Must working memory have the same storage capacity? Must it be equipped with the same kinds of executive operations? We can speculate that creatures with faculties that are very similar to ours are conscious, but what about creatures that attend on the basis of very different kinds of principles? Consider two competing psychological theories of how attention works in us, one of which must be false. Now imagine a possible creature for which the false theory is true. Would that creature be conscious? Who knows?

Things get even trickier when we descend to the level of implementation. Here the method of difference-makers would have us keep algorithms and psychological profiles constant while altering raw materials. For example, we might surgically replace neurons with silicon chips. It is not clear whether this is a nomological possibility. Much will depend on the way we specify our algorithms. If we characterize attention and perception, say, by precisely specifying psychophysical thresholds, similarity spaces and psychometrics of human response, the resulting algorithms (imagine a detailed computer model with lots of equations and variables ranging over time) may be difficult to implement without placing serious constraints on materials. Silicon chips might not work. Let's put that problem to one side. Let's suppose we could do a cell-to-chip replacement while keeping the rules and representations constant. What would we learn about the role of raw materials in the production of consciousness? I think we would learn exactly nothing.

The reason for this is familiar. It harks back to the vast literatures on spectrum inversion and other minds. If we keep other levels constant while performing a cell-to-chip replacement, we will end up with a functional duplicate of a normal person with a normal brain. The victim of this procedure will continue to behave as if she were seeing sunsets and smelling roses. She would report qualitative experiences, and the sensory quality space carved out by her discriminatory capacities and similarity judgments would be isomorphic to the sensory quality space of her pre-surgical self. By all appearances, she would be unchanged. Keeping higher functional levels constant while swapping materials deprives us of any way of detecting whether phenomenology has changed. The measures of experience are behavioural, so the obvious test for the phenomenal impact of matter is going to be utterly uninformative.

This rather obvious point, worn bare by countless discussions, grounds a kind of mysterianism. We can determine whether certain higher-level functional properties are relevant for consciousness. If we disrupt attention, for example, we can be reasonably confident that consciousness is affected. There is some wiggle room for the sceptic here. A sceptic can claim that the phenomenology remains constant during inattention, but loses access to the cognitive centres that are involved in reporting. This scepticism has a kind of stubborn intractability that might ground a mysterian conclusion in its own right. But one can respond by saying that, short of any good reason for thinking that phenomenology continues under inattention, there is no reason to raise the sceptical question. And we do not have an especially good reason to think there is residual phenomenology in this case. Consider the following contrast. Under certain conditions of partially engaged attention, readily perceivable stimuli often go unnoticed. During such episodes, we do not report or act on those stimuli. But, ordinarily, if someone were to bring our attention back around to these stimuli, we would report having been conscious of them, in some sense, all along. Block's (1995) example of hearing a drill in the background while engaged in conversation is an example of this. This case is quite unlike the results that Mack and Rock (1998) report in their experiments on inattentional blindness. When you ask their subjects whether they recall seeing the unexpected stimulus, they say 'no'. This is not what you would expect if, like the drill in the background, they had a phenomenal experience that was simply unavailable for reporting. The fact that subjects have some recall in one case and none in the other is a *prima facie* reason for thinking there is a difference. I venture that phenomenal experiences should be reportable, at least in retrospect, if there is no independent reason for thinking that reporting is impossible.

The alternative hypothesis would be to say that the stimulus used in the inattentional blindness studies was conscious and forgotten. This proposal is doubly flawed: first, it cannot account for why that case differs from the drill case, and, second, it is *ad hoc*. Here's a better explanation. In the drill case, attention is not completely consumed. We are able to monitor the background while engaged in conversation (a good survival skill). In the Mack and Rock studies, attention is more completely consumed through careful experimental manipulation. The difference in attentional demands leads to a difference in consciousness and a difference in access to memory and retrospective report. This explanation has the advantage of taking first-person reports seriously, explaining a difference in performance, and avoiding *ad hoc* postulation. The stimulus is forgotten in the inattentional blindness paradigm because it never gets into working memory, and working memory is necessary for laying down episodic memories (see Prinz, 2001). The stimulus never gets into working memory because consciousness *just is* the process whereby perceptual information becomes available to working memory.

I have been urging that we take first-person reports seriously. Since subjects cannot report seeing the unexpected stimulus in the inattentional blindness experiments, even in retrospect, we assume that they cannot see it. Ostensibly, the same methodological precept could be applied in the cell-to-chip surgery case. Subjects who have undergone this procedure continue to report having phenomenal experiences. Shouldn't we take their first-person reports seriously? Perhaps not. We take phenomenal reports seriously in subjects with organic brains because we have very good reason to think they have phenomenal experiences. After all, their brains are like ours. When you report having an experience, I believe you, and when you insist that you haven't had an experience, I trust you as well, unless I have good reason to think you might have forgotten the experience. All this faith is grounded in the access I have to my own phenomenology and our biological similarities. We cannot make these assumptions when dealing with inorganic brains (e.g., Block, 1978a). With inorganic brains, we must base conclusions about phenomenology on behaviour, and behaviour is not sufficient evidence. To assume otherwise would beg the question against those who say biology is crucial. But we cannot assume that biology is crucial, because that would beg the question against the other side. The problem is that we cannot decide between these options. This, I believe, is the primary reason why we should be mysterians. It has nothing to do with our inability to see the brain in all its glory, as McGinn would have it. Agnosticism follows from the familiar fact that consciousness could alter without affecting behaviour. In ordinary human beings, we can use self-report as a reliable measure, but all bets are off when we consider machines.

V: Objections to the Difference-Maker Problem

The argument for mysterianism trades on my assumptions about cell-to-chip replacement. One might try to combat mysterianism by showing that those assumptions are problematic.

Massive error

One objection owes to Chalmers (1995). In discussing cell-to-chip transformations, he reasons that phenomenology could *not* be lacking in chip-headed people because, if it were, their self reports would be massively erroneous. A theory of consciousness should not allow that someone can be so systematically mistaken about their own mental lives.

I find this objection unconvincing (see also Van Heuveln *et al.*, 1998). What is wrong, I would reply, with attributing massive error? It is conceivable that ordinary human beings are systematically wrong about certain domains (consider probability judgments or error theories of colour), and chip-headed people are far from ordinary. They are victims of invasive surgery, and that surgery is carefully designed to ensure that the functional roles within in their brains remain unaltered. The surgery is set up so that they will continue to make the self-reports that they always have. It is wrong to attribute massive error in cases where error has no explanation, but in the present case it does. If we wired a person's brain so that it caused her to say 'chicken' every time she saw a duck, we would be happy to say she is systematically mistaken in her chicken judgments. If we wire a person's brain to say 'I am having a red experience' every time a ripe tomato is unconsciously perceived, the systematic error would be entirely explicable.

Alternatively, one might respond to Chalmers by arguing that no errors are being made. The chip-headed person's mental terms may come to mean something different from ours. Where 'red experience' refers to a phenomenal state in me, it may come to refer to an unconscious state in my chip-headed companions, because it does not reliably detect anything phenomenal.

The argument from functional analogy

The claim that chip-headed people may lack phenomenology is vulnerable to a second objection. Above, I said that we cannot be sure that they have conscious experiences, because we have only behavioural evidence, and behavioural evidence is insufficient. In actuality, however, we have more than behaviour to go on. We also know that a person who has undergone cell-to-chip surgery has a 'brain' that is functionally organized in the same way as our own. We can base our assessment of such a person's mental life on behaviour *plus* functional isomorphism. And that isomorphism is at an extremely fine level. A chip-headed person has functional organization that is like ours at a level of neurons, not just at the level of folk or scientific psychology. Shouldn't this be enough to block sceptical worries? To say otherwise would be to assume that something about the raw materials of biological brains make a contribution to phenomenology in terms of their intrinsic properties, rather than relational properties. The idea would be that neurons contribute to feels because of something other than the fact that they play certain roles within complex networks. Such intrinsicalism is spooky. It seems to be a consequence of this approach that a single neuron could have phenomenal properties in isolation. Could neurons really have feels in a Petri dish? Sceptics will want to leave open this possibility, but having resisted weird sceptical intuitions elsewhere, shouldn't I do so here?

There are two replies to this kind of objection: one concessive, the other not. The first reply is to concede that a chip-headed person would have phenomenology. This reply entails that artificial experience is possible. But what sort of possibility is this? Since building prosthetic brains is not feasible in the real world, this concession offers no hope to ambitious engineers. One can still claim that, for all we know, building artificially conscious machines is not possible. If we try to simulate a human mind at a higher level of functional organization, our grounds for trusting reports will diminish, and we will return to our initial predicament: one of ignorance. We will have only behaviour and gross functional organization to go on, and *that* does not warrant taking first-person reports seriously. No further test will help, and no appeal to features of the functional duplicates will drive us to certainty, because the dissimilarities between them and us will include both functional differences and physical differences. In sum, the concessive reply says that agnosticism can be averted, but only by building a silicon brain. The really interesting question of machine consciousness concerns the possibility of phenomenology in machines that are quite different from brains at low levels of analysis, and here agnosticism is inescapable.

In a less concessive mood, one might argue that simulated brains are not clear cases of consciousness. Above, I said that one could make this argument only if one defended the view that individual neurons have phenomenal qualities — qualities that might magically ooze from biological structure rather than functional role. Taking this possibility seriously is tantamount, I claimed, to taking seriously the weird possibility of consciousness in a Petri dish. On closer analysis, however, one does not need to take on such weirdness. One could argue that biological materials have the *potential* for consciousness, but require much else besides. Perhaps neurons realize their qualitative potential only when placed in complex networks. The networks would create an environment conducive to consciousness by, say, guaranteeing that the neurons achieve just the right kinds of electrical properties. Alternatively, one might say that an individual quale supervenes on a collection of non-conscious neurons playing a functional role. To assume that conscious states must supervene on conscious building blocks is a fallacy of composition. To assume that conscious states must supervene on biological materials is not. It is simply to vest one's faith in the possibility that biological materials may have properties that are essential ingredients for consciousness. It is taken as given that biological material have some properties that differ from silicon chips, and the claim here is that those properties *might* be important. This possibility is not an example of wild, sceptical standard-raising. It does not commit one to weird possibilities like conscious cells in Petri dishes. The only implication it has is agnosticism about artificial consciousness, and that, I would submit, is not especially weird. I am not suggesting that we can take the necessity of biological ingredients as fact. Instead, I am suggesting that we do not know how to rule out this possibility.

Other minds

This response will seem unsatisfying to those who are persuaded by Turing's (1950) argument against scepticism about machine consciousness. Turing accused sceptics of solipsism. If we encounter a computer that shares much in common with human beings (Turing had behavioural equivalence in mind), then we cannot doubt that it is conscious without also doubting that other human beings are conscious. To doubt that other human beings are conscious is excessively sceptical. It would lead to the repugnant conclusion that I can have knowledge only of my own conscious states. Can we really be agnostic about machine consciousness without being agnostic about other minds?

It is difficult to address this question without getting embroiled in the other minds debate. It used to be popular to think we know about other minds by analogy from our own case. This proposal faces an obvious problem. Induction from a single case is risky. This problem led many authors to reject the argument from analogy in favour of an argument to the best explanation. I think this strategy works for mentality in general, but not necessarily for consciousness. We do not know enough about the causal role of consciousness to know what behaviours can be explained without it. I think we can get a better argument for other minds by combining analogy and best explanation. Here's a quick gloss. I have good reason for being a materialist, by argument to the best explanation. Dualism runs into deep trouble with causation. So I have good reason to think that my conscious states depend on something physical about me. I'm not sure which facts about me are relevant, but I can rule some things out. I know that my own conscious states do not vary in any significant way as a function of, say, what I had for lunch. I remain conscious despite changes in my appearance over time, changes in my beliefs and changes in my neuronal wiring. Consequently, I don't take these things to be essential to consciousness. But other things are harder to rule out. I cannot know whether my consciousness depends on having a brain made out of certain stuff, because my brain has been made out of the same stuff all along. Now consider other minds. I have reason to think that the kinds of physical states that are experienced consciously in me are experienced consciously in others because, as a rough rule, physical states that are alike in all observable respects tend to be alike in respects that are more difficult to observe. When I encounter another human being, I have good reason to think she is conscious because she is like me in some physical respects. The fact that she is unlike me in other physical respects does not matter, because variation along those dimensions does not eliminate consciousness in my own case. With computers, all bets are off because they differ in respects that *may* matter. I do not know for sure that the differences are significant, but I have no way of ruling it out. In other words, there is an asymmetry between the other minds case with humans and the case of computers. When I extend attributions of consciousness from my own case, I can bracket features that I know to be irrelevant and emphasize features that I know to be relevant. Aspects of my biological make-up, however, cannot be put into either category. I'm not sure whether they are relevant or not. With other human beings, this uncertainty is moot because they share my biology. With computers, the uncertainty forces me into a position of agnosticism.

First-person knowledge

I claim we can never know if a chip-headed person has conscious experiences. But surely the chip-headed person would know. From a third-person point of view, mysterianism looms, but the truth is perfectly clear from the first-person perspective. After all, the chip-headed person either has experiences or doesn't have experiences. We can even imagine putting chips in one part of a person's visual field and leaving the rest intact. The person would be able to directly compare what it is like to see the colour of an object placed in a region that has adjacent groups of cellular and microchip receptors. Perhaps a consciousness researcher could do this and come to know whether conscious computers are possible.

This kind of potential for knowledge is not especially useful. A consciousness researcher with a partially silicon brain could not access her direct knowledge in a way that would allow her to draw explicit conclusions about the possibility of conscious machines. Suppose her chip-based receptive fields delivered no qualitative feels. She would still make a judgment that she was having visual experience from these fields even though she would not be having such experiences. The lack of phenomenology would not prevent her from forming false beliefs and coming to false conclusions. And, importantly, she would draw the same conclusions if she *did* have experience from the chip-based receptive fields. Regardless of what she was experiencing, she would come to the same conclusions about the possibility of machine intelligence. Any 'knowledge' she had of what chip receptors are like would do nothing for her ultimate assessment of the question under discussion. Even if she happened to entertain the true thought the she was (un)able to experience anything in her chip receptors, that thought would not count as knowledge because it would not be reliably or inferentially related to the (lack of) experience in question. Reflecting on her own condition, she, too, would be forced to a mysterian position about artificial experience.

One might think this obstacle could be circumvented if we took a chip-headed person and refitted her with biological cells after a brief period with silicon implants. Couldn't that person now recall what it was like to have chip-based receptors and compare the experience to experiences afforded by her biologically reconstituted states? Apparently not. In recalling prior experiences, we presumably do one of two things. Either we call up a propositional or verbal description of a past experience or we reproduce the experience using mental imagery. A former chip-header will have propositional and verbal memories of vivid experiences, because verbal memory would have been unaltered. And if she tried to call up a mental image, she would be forced to activate her new biological cells rather than reactivating their former silicon proxies. She would not be able to directly compare biological and silicon experiences.

Epiphenomenalism

To claim that silicon receptors might do their work without any phenomenal qualities seems to imply that qualia may be epiphenomenal. If cell-to-chip replacement eliminated experience without affecting thought and action, then experience plays no causal role. This is, arguably, an undesirable outcome. It also threatens to call into question the methodology invoked earlier. I said we can try to identify the material correlates of consciousness by seeing what role conscious states play. If qualia are not part of the causal nexus, this methodology is suspect. The possibility of epiphenomenalism doesn't prove that mysterianism is false. Rather, it suggests that one's mysterianism should be more thoroughgoing. If phenomenal qualities are epiphenomenal, we cannot report having them or test for them in any other way. Tests, including first-person introspective judgments, are causal consequences of that for which they test. If epiphenomenalism is an open possibility, then my mysterianism should also extend to take seriously the possibility of dualism. After all, it is far from clear how anything physical could lack causal effects. If I seriously entertain the supposition that cell-to-chip surgery could preserve function without preserving consciousness, then I should also take dualism seriously, because that is what the supposition seems to require.

For brevity, I will restrict myself to one reply. It is not a consequence of my view that qualia are potentially epiphenomenal. That would be a consequence only if epiphenomenalism were entailed by the possibility that chip-headers lack qualia. But no such entailment holds. Causal properties exist at many levels. We can think of a brain as a set of nested mechanisms providing outputs in response to inputs. For each decomposable mechanism, we could, in theory, retain the input/output function by changing the nested mechanism that happens to achieve that function. One can think of individual neurons this way, too. A neuron implements a function from electrical inputs to electrical outputs in virtue of having certain physical properties. One can preserve the function, in theory, by means of different physical properties. Cells and silicon chips work in virtue of different physical properties. So, chip-to-cell surgery changes the causal properties of a brain. If consciousness disappears, it could be in virtue of a change in causal properties. The mechanisms that produce verbal reports are unaltered by the surgery, but that does not mean consciousness is inert. It may operate at a lower level. If this picture is correct, we can talk of consciousness playing a causal role in the genesis of ordinary introspective reports. My report that I am experiencing red is caused by seeing red insofar as the mechanisms that lead from perception to report are implemented by mechanisms that comprise the material basis of experience. By analogy, it is perfectly okay to say of an elevator that its movements are caused by cables and pulleys even though its elevating mechanism might be replaced by a hydraulic system that preserves the input/output mapping from button pushes to floor changes.

I conclude that the possibility of cell-to-chip replacement exposes a serious epistemological problem in consciousness studies. We will never be able to determine what lower levels of analysis are essential for consciousness. The method of difference-makers seems to be the only way to find out what levels matter and, when applied to the level of neurons (or lower), we simply cannot tell whether the transformation makes a difference. There is no way to rule out a biological contribution to consciousness. It does not take a radical form of scepticism to speculate that biology might matter. Neither does it take a radical sceptic to deny that. The result is a radical agnosticism about which of these possibilities is correct.

VI: Concluding Morals

I will conclude with a three brief morals.

First, we must be mysterians about the material basis of consciousness. We can not determine, with any certainty, which low-level properties are essential for consciousness. Changes at the low level that leave no measurable trace in self-report may alter consciousness, but, then again, they may not. When we alter things at higher levels of analysis and find a measurable change in consciousness, we can conclude that something at those levels is making a constitutive contribution. But, when lower levels are altered without causing a measurable change, we cannot infer that nothing at those levels is making a constitutive contribution. This is a fundamental asymmetry brought on by limitations in our ability to measure consciousness.

Second, the science of consciousness is not threatened by this mysterianism. Scientists try to determine the physical basis of consciousness in us. There is good reason to think that attention, perception and access to working memory — all conceived at the level of psychological profile — are all necessary for consciousness. Changes describable at an algorithmic level also have a measurable impact on consciousness. One can generate plausible hypotheses, if not solid proofs, for the importance of certain kinds of rules and representations by studying the processes involved in hierarchical sensory systems, attention and working memory. We may not know exactly which algorithms are necessary, but we have good reason to think that something at this level makes a constitutive contribution. We cannot be sure what, if anything, is necessary for consciousness at the neural level, but we can be sure that certain neuronal processes are sufficient for consciousness. The biological processes that underwrite human perception, attention and working memory have this status. We can continue looking for the neural correlates of human consciousness, and, once we have found them, we will have an account of material conditions that make consciousness possible in us. We are not blind to any properties of the brain, so such an account would not leave anything out.

In sum, we can identify properties that are necessary for consciousness (e.g., attention and perception) and properties that are sufficient (the biological substrates of attention and perception in us), but none that are definitely both necessary and sufficient. This leaves us with a degree of mystery but not one that prevents us from pursuing the science of consciousness just as we are pursuing it.

The third moral derives from the first two. The question of machine consciousness cannot be answered, but that really doesn't matter. If we create an inorganic machine that is isomorphic to a human brain it may lack qualia, for all we know. But building such a machine would probably be impossible. It may even turn out that one needs biological materials to get a machine that has the same dynamics as a human brain. So the project is not worth pursuing. It is worth trying to build machines that mimic human information processing at an algorithmic level. This may be a manageable engineering task, and it can help guide research into how our minds work. Computational models are one of the best ways to test theories of human cognition. If we create an algorithmic analogue of a human mind, we will never know if it is conscious, but the project is independently motivated as a tool for research. So engineers should continue trying to model what they can, and they should stop trying to model what they can't. Mysterianism about machine consciousness, therefore, has little to say about how engineers should go about their business. Engineers shouldn't fool themselves into thinking they can definitely create conscious machines, and they certainly shouldn't feel confident about attaining that goal when they stray too far from the way in which human minds operate. But neither of these points would come as much surprise to anyone in the business. The search for machine consciousness is a perfectly good excuse for building models that can illuminate facts about how minds work, but it is a useless game for those who like to place bets. It is a bit like flipping a coin into the deep sea. We can call heads or tails, but we can't confirm either prediction.[1]

References

Baddely, A.D. and Hitch, G.J. (1977), 'Working memory', in *Recent Advances in Learning and Motivation*, Vol. 8, ed. G. Bower (New York: Academic Press).

Block, N. (1978a), 'Troubles with functionalism', *Minnesota Studies in the Philosophy of Science*, **9**, pp. 261–325.

Block, N. (1978b), 'Reductionism: Philosophical analysis', in *The Encyclopaedia of Bioethics* (London: Macmillan).

Block, N. (1995), 'On a confusion about a function of consciousness', *Behavioural and Brain Sciences*, **18**, pp. 227–87.

[1] I extend thanks to three anonymous referees, to Philip Robbins, and members of the Propositional Attitude Task Force, especially Lee Bowie, Jay Garfield and Laura Sizer. Few readers seem to agree with the core thesis of this paper. I appreciate their efforts to disabuse me of my mysterianism.

Chalmers, D.J. (1995), 'Absent qualia, fading qualia, dancing qualia', in *Conscious Experience*, ed. Metzinger, T. (Thorverton: Imprint Academic).
Churchland, P. S. (1986), *Neurophilosophy* (Cambridge, MA: MIT Press).
Craver, C. (2001). 'Role functions, mechanisms, and hierarchy', *Philosophy of Science*, **68**, pp. 53–74.
Crick, F. and Koch, C. (1990), 'Towards a neurobiological theory of consciousness', *Seminars in the Neurosciences*, **2**, pp. 263–75.
Haugeland, J. (1985), *Artificial Intelligence: The Very Idea* (Cambridge, MA: MIT Press).
Haxby, J.V., Petit, L., Ungerleider, L.G. and Courtney, S.M. (2000), 'Distinguishing the functional roles of multiple regions in distributed neural systems for visual working memory', *Neuroimage*, **11**, pp. 380–91.
Jackendoff, R. (1987), *Consciousness and the Computational Mind* (Cambridge, MA: MIT Press).
Kosslyn, S.M. (1994), *Image and Brain: The Resolution of the Imagery Debate* (Cambridge, MA: MIT Press).
McGinn, C. (1991), *The Problem of Consciousness. Essays Towards a Resolution* (Oxford: Blackwell).
McGinn, C. (1999), *The Mysterious Flame: Conscious Minds in a Material World* (New York: Basic Books).
Mack A. and Rock I. (1998), *Inattentional Blindness* (Cambridge, MA: MIT Press).
Marr, D. (1982), *Vision: A Computational Investigation into the Human Representation and Processing of Visual Information* (New York: W. H. Freeman).
Moran, J. and Desimone, R. (1985), 'Selective attention gates visual processing in the extrastriate cortex', *Science*, **229**, pp. 782–4.
Papineau, D. (2002), *Thinking about Consciousness* (Oxford: Oxford University Press).
Prinz, J.J. (2000), 'A neurofunctional theory of visual consciousness', *Consciousness and Cognition*, **9**, pp. 243–59.
Prinz, J.J. (2001), 'Functionalism, dualism and the neural correlates of consciousness', in *Philosophy and the Neurosciences: A Reader*, ed. W. Bechtel, P. Mandik, J. Mundale and R. Stufflebeam (Oxford: Blackwell).
Prinz, J.J. (2002). *Furnishing the Mind: Concepts and Their Perceptual Basis* (Cambridge, MA: MIT Press).
Rees, G. (2001), 'Neuroimaging of visual awareness in patients and normal subjects', *Current Opinion in Neurobiology*, **11**, pp. 150–6.
Rosenthal, D.M. (1986), 'Two concepts of consciousness', *Philosophical Studies*, **49**, pp. 329–59.
Searle, J.R. (1980), 'Minds, brains, and programs', *Behavioural and Brain Sciences*, **3**, pp. 417–57.
Searle, J.R. (1992), T*he Rediscovery of Mind* (Cambridge, MA: MIT Press).
Tootell R.B.H., Reppas J.B., Dale A.M., Look R.B., Sereno M.I., Malach R., Brady T.J. and Rosen B.R. (1995), 'Visual motion aftereffect in human cortical area MT revealed by functional magnetic resonance imaging', *Nature*, **375**, pp. 139–41.
Turing, A.M. (1950), 'Computing machinery and intelligence,' *Mind*, **59**, pp. 433–60.
Van Heuveln, B., Dietrich, E. and Oshima, M. (1998), 'Let's dance! The equivocation in Chalmers' dancing qualia argument', *Minds and Machines*, **8**, pp. 237–49.
Zeki, S. (1993), *A Vision of the Brain* (London: Blackwell).

Aaron Sloman and Ron Chrisley

Virtual Machines and Consciousness

Replication or even modelling of consciousness in machines requires some clarifications and refinements of our concept of consciousness. Design of, construction of, and interaction with artificial systems can itself assist in this conceptual development. We start with the tentative hypothesis that although the word 'consciousness' has no well-defined meaning, it is used to refer to aspects of human and animal information processing. We then argue that we can enhance our understanding of what these aspects might be by designing and building virtual-machine architectures capturing various features of consciousness. This activity may in turn nurture the development of our concepts of consciousness, showing how an analysis based on information processing virtual machines answers old philosophical puzzles as well enriching empirical theories. This process of developing and testing ideas by developing and testing designs leads to gradual refinement of many of our pre-theoretical concepts of mind, showing how they can be construed as implicitly 'architecture-based' concepts. Understanding how human-like robots with appropriate architectures are likely to feel puzzled about qualia may help us resolve those puzzles. The concept of 'qualia' turns out to be an 'architecture-based' concept, while individual qualia concepts are 'architecture-driven'.

I: Introduction

The study of consciousness, a long-established part of philosophy of mind and of metaphysics, was banished from science for many years, but has recently re-entered (some would say by a back door that should have been kept locked). Most AI researchers ignore the topic, though people discussing the scope and limits of AI do not. We claim that much of the discussion of consciousness is confused because what is being referred to is not clear. That is partly because 'consciousness' is a *cluster concept*, as explained below.

Progress in the study, or modelling, of consciousness requires some clarifications and refinements of our concepts. Fortunately, design of, construction of, and interaction with artificial systems can itself assist in this conceptual development. In particular, we start with the tentative hypothesis that although the word 'consciousness' has no well-defined meaning, it is used to refer to a cluster of aspects of information processing in humans and other animals. On that basis we can enhance our understanding of what these aspects might be by designing, building, analysing, and experimenting with virtual-machine architectures which attempt to elaborate the hypothesis. This activity may in turn nurture the development of our concepts of consciousness, along with a host of related concepts, such as 'experiencing', 'feeling', 'perceiving', 'believing', 'wanting', 'enjoying', 'remembering', 'noticing' and 'learning', helping us to see them as dependent on an implicit theory of minds as information processing virtual machines. On this basis we can find new answers to old philosophical puzzles as well as enriching our empirical theories. We expect this process to lead to gradual refinement and extensions of many of our pre-theoretical concepts of mind as 'architecture-based' concepts, partly mirroring the development of other pre-scientific concepts under the influence of scientific advances (Sloman, 2002). The result, it is hoped, is that the successor concepts will be free of the many conundra (such as the apparent possibility of zombies) which plague our current, inchoate concept of consciousness.

Specifically, we hope to explain how an interest in questions about consciousness in general and *qualia* in particular arises naturally in intelligent machines with a certain sort of architecture that includes a certain sort of 'meta-management' layer. Explaining the possibility (or near-inevitability) of such developments illustrates the notion of 'architecture-driven' concepts (concepts likely to be generated within an architecture) and gives insightful new explanations of human philosophical questions, and confusions, about consciousness.

We emphasise the importance of the notion of a virtual machine architecture and use that as the basis of a notion of *virtual machine functionalism* (VMF) which is immune to the common attacks on more conventional functionalist analyses of mental concepts which require all internal states and processes to be closely linked to possible input–output relations of the whole system. We propose that science, engineering and philosophy should advance simultaneously, and offer a first-draft, general architectural schema for agent architectures, which provides a useful framework for long-term AI research both on human-like systems and models of other animals, and may also inspire new developments in neuropsychology, and a new understanding of the evolution of mind as well as advancing philosophy.

II: Confused Concepts of Consciousness

Before considering the possibility and details of machine consciousness, we might wonder: what do we mean by 'consciousness'? Let's start with some questions:

- Is a fish conscious?
- Is a fly conscious of the fly-swatter zooming down at it?
- Is a new-born baby conscious (when not asleep)?
- Are you conscious when you are dreaming? Or sleep-walking?
- Is the file protection system in an operating system conscious of attempts to violate access permissions?
- Is a soccer-playing robot conscious? Can it be conscious of an opportunity to shoot?

Not only do different people give different answers to these and similar questions, but it seems that what they understand consciousness to be varies with the question. A central (though unoriginal) motif of this paper is that our current situation with respect to understanding consciousness (and many other mental phenomena, e.g. learning, emotions, beliefs) is similar to the situation described in 'The Parable of the Blind Men and the Elephant'.[1] Six blind men encounter an elephant. Each feels a different part, and infers from the properties of the portion encountered the nature of the whole (one feels the tusk and concludes that he has encountered a spear, another feels the trunk and deduces that he has met a snake, etc.). It is often suggested that we are in the same position with respect to consciousness: different (even incompatible) theories may be derived from correct, but incomplete, views of reality.

This is partly the result of (or partly the cause of) the fact that the concept 'consciousness' is not 'well-behaved', in several ways. For one thing, it is a *cluster concept*,[2] in that it refers to a collection of loosely-related and ill-defined phenomena. This is worse than merely being a *vague* concept (e.g. 'large, 'yellow'), for which a boundary cannot be precisely specified along some continuum. It is also worse than being a *mongrel concept* (Block, 1995), which merely confuses a collection of concepts that are individually supposed to be well-defined (e.g., phenomenal consciousness' and 'access consciousness'). It is true that even in the case of vague, mongrel, and even well-behaved concepts, people often disagree on whether or how the concept is to be applied in a particular case. But what is particularly problematic about the concept 'consciousness' is that such disagreement is not merely empirical, but (also) deeply conceptual, since it is often the case that disputants cannot even agree on what sorts of evidence would settle the question. Finally, the cluster concept nature of the concept 'consciousness' is exacerbated by the fact that it is context-sensitive: *which* ill-defined collection of phenomena is being gestured towards itself changes with context, e.g. when we think about consciousness in different animals, or in human infants.

This indeterminacy generates unresolvable disputes about difficult cases. However, some cluster concepts can be refined by basing more determinate

[1] John Godfrey Saxe, 1816–1887; see, e.g., http://www.wvu.edu/~lawfac/jelkins/lp-2001/saxe.html

[2] The phrase 'cluster concept' seems to have been coined by D. Gasking and has been in intermittent use since the mid-twentieth century. Closely related notions are 'family resemblance concept' (Wittgenstein, 1953), 'open texture' (Waismann, 1965), and Minsky's notion of a 'suitcase concept' used in his draft book on *Emotions* online at http://www.media.mit.edu/~minsky/. Compare Ch. XI in Cohen (1962).

variants of them on our architectural theories, as happened to concepts of kinds of matter when we learnt about the architecture of matter. Indeterminate concepts bifurcate into more precise variants (Sloman, 2002). We'll illustrate this below.

1. Evidence for confusion, and partial diagnosis

Our blind (or short-sighted) groping, together with our struggles to treat an indeterminate cluster concept as if it were well defined, have resulted in an astonishing lack of consensus about consciousness, as illustrated in table 1.

Table 1. A Babel of views on the nature of consciousness

Some say consciousness is . . .	**While others say consciousness is . . .**
Absent when you are asleep	Present when you dream
Absent when you are sleep-walking	Present when you are sleep-walking
Essential for processes to be mental	Not required (there are unconscious mental processes)
Able to cause human decisions and action	Epiphenomenal (causally inefficacious)
Independent of physical matter (i.e. disembodied minds are possible)	A special kind of stuff somehow produced by physical stuff
Just a collection of behavioural dispositions	Just a collection of brain states and processes, or a neutral reality which has both physical and mental aspects
Just a myth invented by philosophers, best ignored	Something to do with talking to yourself
Something you either have or don't have	A matter of degree (of something or other)
Impossible without a public (human) language	Present in animals without language
Present only in humans	Present in all animals to some degree
Located in specific regions or processes in brains	Non-localisable; talk about a location for consciousness is a 'category mistake'
Necessarily correlated with specific neural events	Multiply realisable, and therefore need not have fixed neural correlates
Not realisable in a machine	Realisable in a machine that is (behaviourally; functionally) indistinguishable from us
Possibly absent in something (behaviourally; functionally) indistinguishable from us (zombies)	Necessarily possessed by something which has the same information processing capabilities as humans

Some people offer putative definitions of 'consciousness', for instance defining it as 'self-awareness', 'what it is like to be something', 'experience', 'being the subject of seeming' or 'having somebody home', despite the fact that nothing is achieved by defining one obscure expression in terms of another. We need to find a way to step outside the narrow debating arenas to get a bigger picture. Definitions are fine when based on clear prior concepts. Otherwise we need an alternative approach to expose the conceptual terrain. Hopefully then we'll then see all the sub-pictures at which myopic debaters peer, and understand why their descriptions are at best only part of the truth.

In order to see the bigger picture, it will help to ask why there is a Babel of views in the first place. We suggest that discussion of consciousness is confused for several reasons, explained further below:

- Some people focus on one case: normal adult (academic?) humans, whilst others investigate a wider range of cases, including people with brain damage, infants, and other animals.
- Many thinkers operate with limited ideas about possible types of machines (due to deficiencies in our educational system).
- There is especially a lack of understanding about virtual, information-processing machines resulting in ignorance or confusion about entities, events, processes and states that can exist in such machines.
- Many people are victims of the illusion of 'direct access' to the nature of consciousness, discussed below.
- Some people want consciousness to be inexplicable by science, while others assume that it is a biological phenomenon eventually to be explained by biological mechanisms. (We aim to show how this might be done.)

2. Introspection can be deceptive

A Golden Rule for studying consciousness: Do not assume that you can grasp the full nature of consciousness *simply* by looking inside yourself, however long, however carefully, however analytically.

Introspection, focusing attention inwardly on one's own mental states and processes, is merely one of many types of perception. Like other forms of perception it provides only information that the perceptual mechanism is able to provide! Compare staring carefully at trees, rocks, clouds, stars and animals hoping to discover the nature of matter. At best you learn about a subset of what needs to be explained. Introspection can also give incorrect information, for instance convincing some people that their decisions have a kind of freedom that is incompatible with physical causation, or giving the impression that their visual field is filled with uniformly detailed information in a wide solid angle, whereas simple, familiar experiments show that in the region of the blind-spot there is an undetectable information gap. People can be unaware even of their own strong emotions, such as jealousy, infatuation and anger. Introspection can deceive people into thinking that they understand the notion of two distant events being simultaneous, even when they don't: simultaneity can be experienced directly, but Einstein showed that that did not produce full understanding of it.

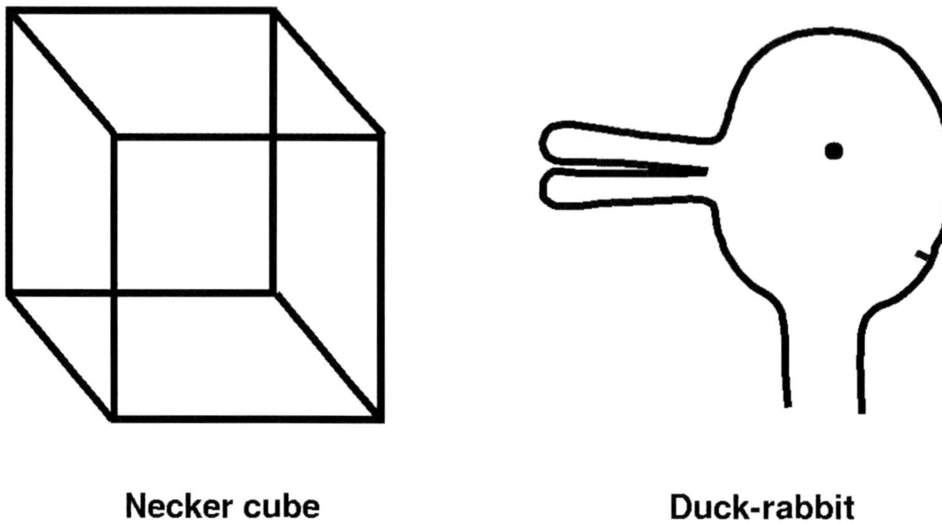

Figure 1. The Necker cube and the duck-rabbit: both are visually ambiguous, leading to two percepts. Describing what happens when they 'flip' shows that one involves only geometrical concepts whereas the other is more abstract and subtle.

3. Introspection can be useful for science

However, introspection is not useless: on the contrary, introspectively analysing the differences between the (probably familiar) visual flips in the two pictures in figure 1 helps to identify the need for a multi-layered perceptual system, described below. When the Necker cube 'flips', all the changes are geometric. They can be described in terms of relative distance and orientation of edges, faces and vertices. When the duck-rabbit 'flips', the geometry does not change, though the functional interpretation of the parts changes (e.g., 'bill' into 'ears'). More subtle features also change, attributable only to animate entities. For example, 'Looking left', or 'looking right'. A cube cannot look left or right. What does it *mean* to say that you 'see the rabbit facing to the right'? Perhaps it involves seeing the rabbit as a *potential mover*, more likely to move right than left. Or seeing it as a *potential perceiver*, gaining information from the right. What does categorising another animal as a perceiver involve? How does it differ from categorising something as having a certain shape?[3] We return to the multiplicity of perception, in explaining the H-CogAff architecture, in section V.5.

These introspections remind us that the experience of seeing has hidden richness, involving a large collection of unrealised, un-activated, but potentially activatable capabilities, whose presence is not immediately obvious. Can we say more about what they are? One way is to learn from psychologists and brain scientists about the many bizarre ways that these capabilities can go wrong. But we can also learn new ways of looking at old experiences: For example, how exactly do you experience an empty space, as in figure 2?

[3] Compare the discussion of 'Seeing as' in Part 2 section xi of Wittgenstein (1953).

Figure 2. The final frontier?

Humans (e.g. painters, creators of animated cartoons, etc.) can experience an empty space as full of possibilities for shapes, colours, textures, and processes in which they change. How? Can any other animal? Current AI vision systems cannot; what sort of machine could? An experience is constituted partly by the collection of implicitly understood *possibilities for change* inherent in that experience. This is closely related to Gibson's 'affordance' theory (Gibson, 1986; Sloman, 1989;1996) discussed below.[4]

> **The Iceberg conjecture:** Consciousness as we know it is *necessarily* the tip of an iceberg of information processing that is mostly totally inaccessible to consciousness. We do not experience what it is to experience something.

The examples show that when we have experiences there may be a lot going on of which we are normally completely unaware, though we can learn to notice some of it by attending to unobvious aspects of experiences, e.g. unnoticed similarities and differences. So we use introspection to discover things about the phenomenology of experience — contributing both to the catalogue of what needs to be explained and to specifications of the internal states and processes required in a human-like machine.[5]

4. Beyond introspection

Although introspection is useful, we must do more than gaze at our internal navels; we need to collect far more data-points to be explained, e.g. concerning:

- the varieties of tasks for which different sorts of experiences are appropriate — e.g. what sorts of experiences support accurate grasping movements, obstacle avoidance, dismantling and re-assembly of a clock, avoiding a predator, catching fast moving prey, noticing that you are about to say something inconsistent, being puzzled about something, etc.;

[4] Compare Wittgenstein: 'The substratum of this experience is the mastery of a technique' (op. cit.).

[5] A more robust argument for the Iceberg conjecture, especially in its strong modal form, above, cannot be given here; one can find some support for it in Wittgenstein (1922, cf 2.172: 'The picture, however, cannot represent its form of representation; it shows it forth.') and in Smith (1996, p. 303).

- individual differences e.g. experiences at various stages of human development;
- culture-based differences in mental phenomena, e.g. *feeling sinful* is possible in some cultures but not others, and experiencing this text as meaningful depends on cultural training;
- surprising phenomena demonstrated in psychological experiments (e.g. change blindness) and surprising effects of brain damage or disease, including the fragmentation produced by severing the corpus callosum, or bizarre forms of denial;
- similarities and differences between the experiences of different species;
- stages and trends in the evolution of various mental phenomena.

Insofar as different organisms, or children, or people with various sorts of brain damage or disease, have different kinds of mental machinery, different information processing architectures, the types of experiences possible for them will be different. Even if a lion lies down with a lamb, their visual experiences when gazing at the same scene may be different because of the affordances acquired in their evolutionary history.

Understanding our own case involves seeing how we fit into the total picture of biological evolution and its products, including other possible systems on other planets, and also in future robots. It is important to pursue this idea without assuming that states of consciousness of other animals can be expressed in our language. E.g. 'the deer sees the lion approaching' may be inappropriate if it implies that the deer uses something like our concepts of 'lion' and 'approaching'. We must allow for the existence of forms of experience that we cannot describe.[6]

However, 'ontological blindness' can limit the data we notice: we may lack the ability to perceive or conceive of some aspects of what needs to be explained, like people who understand what velocity is but do not grasp that a moving object can have an instantaneous acceleration and that acceleration can decrease while velocity is increasing. *We do not yet have the concepts necessary for fully understanding what the problem of consciousness is.* So effective collection of data to be explained often requires us to refine our existing concepts and develop new ones — an activity which can be facilitated by collecting and attempting to assimilate and explain new data, using new explanatory theories, which sometimes direct us to previously unnoticed phenomena.

We need deeper, richer forms of explanatory theories able to accommodate *all* the data-points, many of which are qualitative (e.g. structures and relationships and changes therein) not quantitative (i.e. not just statistical regularities or functional relationships) and are mostly concerned with what *can* happen or can be done, rather than with predictive laws or correlations.

The language of physics (mainly equations) is not as well suited to describing these realms of possibility as the languages of logic, discrete mathematics,

[6] That we cannot *express* such experiential contents in language does not preclude us from using language (or some other tool) to refer to or specify such contents; see Chrisley (1995).

formal linguistics (grammars of various kinds) and the languages of computer scientists, software engineers and AI theorists. The latter are languages for specifying and explaining the behaviour of information processing machines. We do not claim that *existing* specialist languages and ontologies are sufficient for describing and explaining mental phenomena, though they add useful extensions to pre-theoretic languages.

5. *Virtual machines and consciousness*

It is widely accepted that biological organisms use information about the environment in selecting actions. Often information about their own state is also used: deciding whether to move towards visible food or visible water on the basis of current needs. Consider this bolder claim:

> **Basic working assumption:** The phenomena labelled 'conscious' involve no magic; they result from the operation of very complex biological information processing machines which we do not yet understand.

Although the first part is uncontentious to anyone of a naturalist bent, the latter half is, of course, notoriously controversial. The rest of this paper attempts to defend it, though a full justification requires further research of the sort we propose.

All biological information processing is based on physical mechanisms. That does not imply that information processing states and processes are physical states and processes in the sense of being best described using the concepts of the physical sciences (physics, chemistry, astronomy, etc.). Many things that are produced by or realised in physical resources are non-physical in this sense, e.g. poverty, legal obligations, war, etc. So we can expect all forms of consciousness to be *based on*, or *realised in*, physical mechanisms, but not necessarily to be physical in the sense of being *describable* in the language of the physical sciences.

One way to make progress is to complement research on physical and physiological mechanisms by (temporarily) ignoring many of the physical differences between systems and focus on higher level, more abstract commonalities. For that we need to talk about what software engineers would call the virtual information processing machines *implemented in* those physical machines. Philosophers are more likely to say the former are *supervenient on* the latter (Kim, 1998): both have a partial, incomplete, view of the same relation. It is important that, despite the terminology, virtual machines are *real* machines, insofar as they can affect and be affected by the physical environment. Decision-making in a virtual machine can be used to control a chemical factory for example.

There are many families of concepts, or 'levels' on which we can think about reality (see figure 3 for some examples). At all levels there are objects, properties, relations, structures, mechanisms, states, events, processes and causal interactions (e.g. poverty can cause crime). But we are not advocating a 'promiscuous' pluralism; the world is one in at least this sense: all non-physical levels are ultimately implemented in physical mechanisms, even if they are not

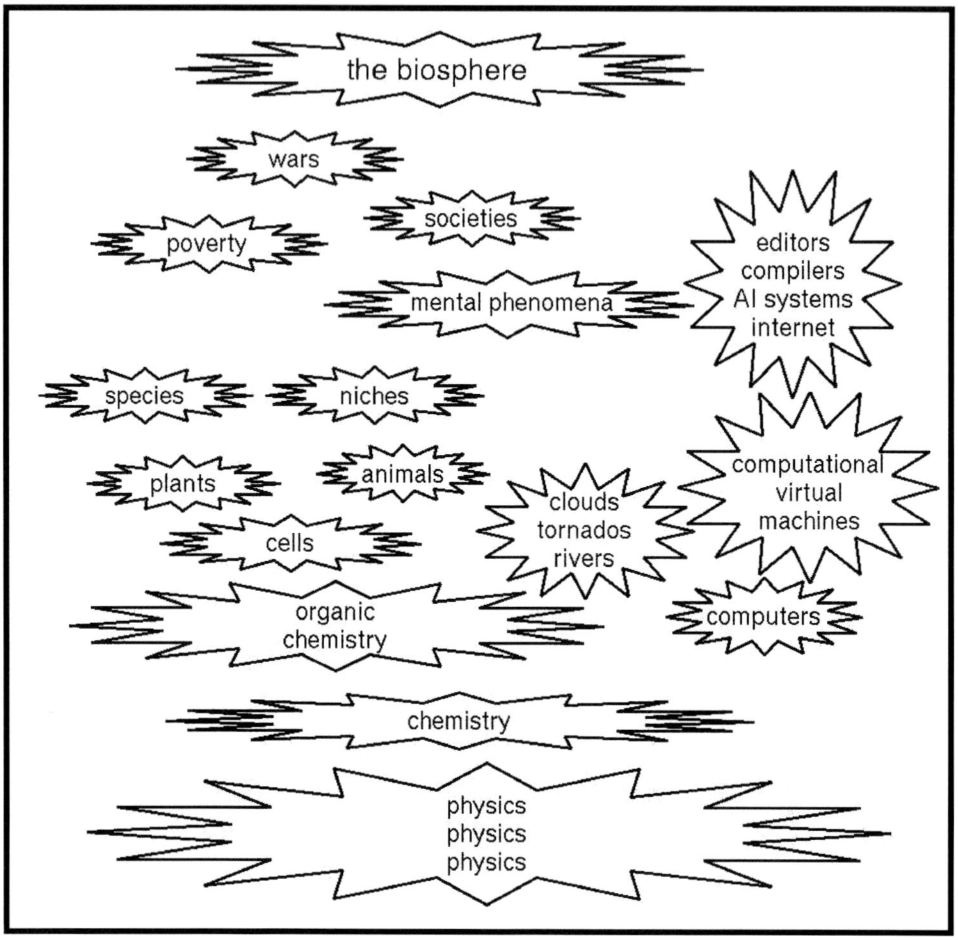

Figure 3. Various levels of reality, most of which are non-physical

definable in the language of physics. The history of human thought and culture shows not only that we are able to make good use of ontologies that are not definable in terms of those of the physical sciences, but that we cannot cope without doing so. Moreover, nobody knows how many levels of virtual machines physicists themselves will eventually discover.

So when we talk about information processing virtual machines, this is no more mysterious than our commonplace thinking about social, economic, and political states and processes and causal interactions between them.

III: The Concept of an 'Information Processor'

Successful organisms are information processors. This is because organisms, unlike rocks, mountains, planets and galaxies, typically require action to survive, and actions must be selected and initiated under certain conditions. The conditions do not directly trigger the actions (as kicking a ball triggers its motion): rather organisms have to initiate actions using *internal* energy. Therefore appropriate and timely selection and initiation of action requires, at a minimum, information about whether the suitability conditions obtain.

1. What is information?

Like many biologists, software engineers, news reporters, propaganda agencies and social scientists, we use 'information' not in the technical sense of Shannon and Weaver, but in the sense in which information can be true or false, or can more or less accurately fit some situation, and in which one item of information can be inconsistent with another, or can be derived from another, or may be more general or more specific than another. This is *semantic* information, involving reference to something. None of this implies that the information is expressed or encoded in any particular form, such as sentences or pictures or bit-patterns, neural states, or that it is necessarily communicated between organisms, as opposed to being acquired or used within an organism.[7]

2. We don't need to define our terms

It is important to resist premature demands for a strict definition of 'information'. Compare 'energy' — that concept has grown much since the time of Newton, and now covers forms of energy beyond his dreams. Did he understand what energy is? Yes, but only partly. Instead of defining 'information' we need to analyse the following:

- the variety of *types* of information there are,
- the kinds of *forms* they can take,
- the kinds of *relations* that can hold between information items,
- the means of *acquiring* information,
- the means of *manipulating* information,
- the means of *storing* information,
- the means of *communicating* information,
- the *purposes* for which information can be used,
- the variety of *ways of using* information.

As we learn more about such things, our concept of 'information' grows deeper and richer, just as our concept of 'energy' grew deeper and richer when we learnt about radiant energy, chemical energy and mass energy. It is a part of our 'interactive conceptual refinement' methodology that although we start with tentative 'implicit' definitions, we expect future developments (including, but not limited to, new empirical data) to compell us to revise them. Neither theory nor data has the final say; rather, both are held in a constructive dialectical opposition (Chrisley, 2000).

---

[7] We have no space to rebut the argument in Rose (1993) that only computers, not animals or brains, are information processors, and the 'opposite' argument of Maturana and Varela summarised in Boden (2000) according to which *only* humans process information, namely when they communicate via external messages. Further discussion on this topic can be found at http://www.cs.bham.ac.uk/research/cogaff/

Like many deep concepts in science, 'information' is *implicitly* defined by its role in our theories and our designs for working systems. To illustrate this point, we offer some examples[8] of processes involving information in organisms or machines:

- external or internal actions triggered by information,
- segmenting, clustering labelling components within a structure (i.e. parsing),
- trying to derive new information from old (e.g. what caused this? what else is there? what might happen next? can I benefit from this?),
- storing information for future use (and possibly modifying it later),
- considering and comparing alternative plans, descriptions or explanations,
- interpreting information as instructions and obeying them, e.g. carrying out a plan,
- observing the above processes and deriving new information thereby (self-monitoring, self-evaluation, meta-management),
- communicating information to others (or to oneself later),
- checking information for consistency.

Although most of these processes do not involve *self*-consciousness, they do involve a kind of awareness, or sentience of something, in the sense of having, or being able to acquire, information about something. Even a housefly has that minimal sort of consciousness. Later we analyse richer varieties of consciousness.

3. Varieties of information contents

Depending on its needs and its capabilities, an organism may use information about:

- density gradients of nutrients in the primaeval soup,
- the presence of noxious entities,
- where the gap is in a barrier,
- precise locations of branches in a tree as it flies through them,
- how much of its nest it has built so far,
- which part of the nest should be extended next,
- where a potential mate is,
- something that might eat it,
- something it might eat,
- whether that overhanging rock is likely to fall,
- whether another organism is likely to attack or run,
- how to achieve or avoid various states,
- how it thought about that last problem,
- whether its thinking is making progress.

Information contents can vary in several dimensions, for instance whether the information is localised (seeing the colour of a dot) or more wholistic (seeing a

[8] This list, and the other lists we present, are meant only to be illustrative, not exhaustive.

tree waving in the breeze), whether it involves only geometric and physical properties (seeing a blue cube) or more abstract properties (seeing a route through shrubbery, seeing someone as angry), whether it refers to something that cannot be directly experienced (electrons, genes, cosmic radiation), whether it refers to something which itself refers to or depicts something else, and so on. Other dimensions of variation include the structure, whether the information involves use of concepts, and objectivity of what is referred to and how it is referred to. (Discussed below in connection with qualia.)

4. Information processing and architecture

What an organism or machine can do with information depends on its architecture. An architecture includes, among other things:

- forms of representation, i.e. ways of storing and manipulating information (Peterson, 1996),
- algorithms,
- concurrently active sub-systems, with different functional roles,
- connections between and causal interactions between sub-systems.

Some architectures *develop* i.e. they change themselves over time so that the components and links within the architecture change. A child's mind and a multi-user, multi-purpose computer operating system in which new interacting processes are spawned or old ones killed, are both examples of changing architectures.

5. What is a machine?

We understand a machine to be a complex whole made of interacting components whose capabilities combine to enable the whole machine to do things. Some of the components may themselves be machines in this sense. There are at least three types of machines:

- Matter manipulating machines: *Diggers, drills, cranes, cookers . . .*
- Energy manipulating machines: *Diggers, drills, cranes, cookers, transformers, steam engines . . .*
- Information manipulating machines: *Thermostats, controllers, most organisms, operating systems, compilers, business organisations, governments . . .*

6. Information processing virtual machines

We are concerned with the third class, the information processing machines. Information-manipulation is not restricted to physical machines, e.g. made of blood, meat, wires, transistors, etc. *Virtual* machines (VM) can also do it. These contain *abstract* non-physical entities, like words, sentences, numbers, bit-patterns, trees, procedures, rules, etc., and the causal laws that summarise their operation are not the same as the laws of the physical sciences.

It may be true of a chess virtual machine that whenever it detects that its king is threatened it attempts to make a defensive move or a counter-attack, but that is not a law of physics. Without changing any laws of physics, the virtual machine can be altered to play in 'teaching mode' for beginners, so that the generalisation no longer holds. The predictability of a chess virtual machine depends in a complicated way on the fact that the components in which it is implemented obey the laws of physics, though the very same sort of virtual machine could be implemented in different components with different behaviours (e.g. in valves instead of transistors).

The laws that govern the VM are not derivable by pure mathematics or pure logic from a physical description of the components and the physical laws governing their behaviour. 'Bridging laws' relating the states and processes in the virtual machine and those in the physical machine are needed. These cannot be proved by logic alone because the concepts required to define a chess virtual machine (e.g. 'queen', 'check', 'win', 'capture') are not *explicitly definable* in terms of those of physics. Neither are bridging laws empirical, since when an implementation is understood, the connection is seen to be necessary because of subtle and complex structural relations between the physical machine and the virtual machine, despite the different ontologies involved.[9]

7. Evolution of information processing architectures

Exploring the full range of designs for behaving systems requires knowledge of a wide range of techniques for constructing virtual machines of various sorts. Many clues come from living things, since evolution 'discovered' and used myriad mechanisms long before human engineers and scientists existed. The clues are not obvious, however: paleontology shows development of physiology and provides weak evidence about behavioural capabilities, but virtual machines leave no fossils. Surviving systems give clues, however. Some information processing capabilities (e.g. most of those in microbes and insects) are evolutionarily very old, others relatively new (e.g. the ability to learn to read, design machinery, do mathematics, or think about your thought processes.) The latter occur in relatively few species. Perceptual mechanisms that evolved at different times provide very different sorts of information about the environment. An amoeba cannot acquire or use information about the state of mind of a human, though a dog can to some extent. Most organisms, including humans, contain both old and new sub-systems performing different, possibly overlapping, sometimes competing tasks. We need to understand how the new mechanisms associated with human consciousness differ from, how they are built on, and how they interact with the older mechanisms.

In order to understand consciousness as a biological phenomenon, we need to understand the variety of biological information processing architectures and the

[9] These are controversial comments, discussed at greater length, though possibly not yet conclusively, in Sloman & Scheutz (2001). Scheutz (1999) argues that the existence of virtual machines with the required structure can be derived mathematically from a description of the physical machine by abstracting from details of the physical machine.

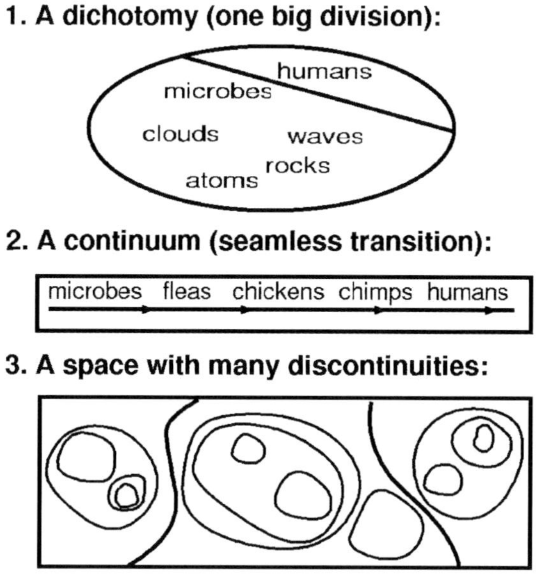

Figure 4. Models of conceptual spaces. It is often assumed that the only alternative to a dichotomy (conscious/non-conscious) is a continuum of cases with only differences of degree. There is a third alternative.

states and processes they can support. There's no need to assume that a unique, correct architecture for consciousness exists; the belief that there is amounts to believing the conscious/non-conscious distinction is a dichotomy, as in figure 4. Many assume that the only alternative to a dichotomy is a continuum, in which all differences are differences of degree and all boundaries are arbitrary, or culturally determined. This ignores conceptual spaces that have many discontinuities. Examples are the space of possible designs and the space of requirements for designs, i.e. 'niche-space' (Sloman, 2000). Biological changes, being based on molecular structures, are inherently discontinuous. Many of the changes that might be made to a system (by evolution or learning or self-modification) are discontinuous; some examples are: duplicating a structure, adding a new connection between existing structures, replacing a component with another, extending a plan or adding a new control mechanism.

We don't know what sorts of evolutionary changes account for the facts that humans, unlike all (or most) other animals, can use subjunctive grammatical forms, can think about the relation between mind and body, can learn predicate calculus and modal logic, can see the structural correspondence between four rows of five dots and five rows of four dots, and so on. We don't know how many of the evolutionary changes leading to human minds occurred separately in other animals (Hauser, 2001). But it is unlikely to have been either a single massive discontinuity, or a completely smooth progression. Below we explore some interesting discontinuities (some of which may result from smoother changes at lower levels), but first a methodological detour.

IV: Varieties of Functionalism

What we are proposing is a new kind of functionalist analysis of mental concepts. Functionalism is fairly popular among philosophers, though there are a number of standard objections to it. We claim that these objections can be avoided by basing our analyses on *virtual machine functionalism*.

1. Atomic state functionalism

Most philosophers and cognitive scientists write as if 'functionalism' were a well-defined, generally understood concept. E.g. Block (1996) writes (regarding a mental state S1):

> According to functionalism, the nature of a mental state is just like the nature of an automaton state: constituted by its relations to other states and to inputs and outputs. All there is to S1 is that being in it and getting a 1 input results in such and such, etc. According to functionalism, all there is to being in pain is that it disposes you to say 'ouch', wonder whether you are ill, it distracts you, etc.

This summary has (at least) two different interpretations, where the second has two sub-cases described in subsection 3 below. On the first reading, which we'll call *atomic state functionalism*, an entity A can have only one, indivisible, mental state at a time, and each state is completely characterised by its role in a state-transition network in which current input to A plus current state of A determine the immediately following output of A and next state of A. This seems to be the most wide-spread interpretation of functionalism, and it is probably what Block intended, since he led up to it using examples of finite-state automata which can have only one state at a time. These could not be states like human hunger, thirst, puzzlement or anger, since these can coexist and start and stop independently.

On the second interpretation of Block's summary, A can have several coexisting, independently varying, interacting mental states. It is possible that Block did not realise that his examples, as ordinarily understood, were of this kind: for instance, the same pain can both make you wonder whether you are ill and distract you from a task, so that having a pain, wondering whether you are ill, having the desire or intention to do the task, and being distracted from it are four coexisting states which need not start and end at the same time. If the pain subsides, you may continue to wonder whether you are ill, and while working on the task (e.g. digging the garden) you might form a new intention to see a doctor later. Coexistence of interacting sub-states is a feature of how people normally view mental states, for instance when they talk about conflicting desires or attitudes.

2. Virtual machine functionalism

If a mind includes many enduring coexisting, independently varying, causally interacting, states and processes then it is a complex mechanism, though not a physical mechanism, since the parts that interact are not physical parts but things like desires, memories, percepts, beliefs, attitudes, pains, etc. This is close to the notion of virtual machine explained earlier, so we'll call this interpretation of Block's summary, allowing multiple, concurrently active, interacting mental states, *virtual machine functionalism* (VMF).

VMF allows that an individual A can have many mental sub-states at any time. Each sub-state S of A will typically depend in part on some sub-mechanism or sub-system in A which produces it (e.g. perception sub-system, action sub-system, long term memory, short term memory, current store of goals and plans, reasoning sub-system, etc.), though exactly which states and sub-systems can coexist in humans is an empirical question. We need both sub-states and sub- systems because the sub-systems endure while the states they produce change: e.g. the visual sub-system persists while what is seen changes. We do not assume that functionally distinct sub-systems necessarily map onto physically separable sub-systems.

Each sub-state S is defined by its causal relationships to other sub-states (and therefore to the sub-systems that produce them) and, in some cases, its causal relations to the environment. (e.g., if S is influenced by sensors or if it can influence motors or muscles) Then a particular type of sub-state S (believing something, wanting something, trying to solve a problem, enjoying something, having certain concepts, etc.) will be defined partly by the kinds of causal influences that state can receive from the environment or from other sub-states and partly by the kinds of causal influences it can have on other sub-states and the environment. The causal influences surrounding a sub-state, or its own internal processes, may cause that state to be replaced by another. For instance A's worry may cause A to remember a previous occasion which causes A to decide that the pain does not indicate illness: so the state of worry causes its own demise. The state of thinking about a problem or imagining how a tune goes will naturally progress through different stages. Each stage in the process will have different causal properties, as Ryle (1949) noted.

Thus, according to VMF, each sub-state S (or persisting process) of agent A may be characterised by a large collection of conditionals of forms like:

> *If an individual A is in sub-state S, and simultaneously in various other sub-states (where each sub-state is produced by some enduring sub-system), then if the sub-system of A producing S receives inputs I_1, I_2, ... from other sub-systems or from the environment of A, and if sub-states S_k, S_l, ... exist then S will cause output O_1 to the sub-system concerned with state S_k and output O_2 to the sub-system concerned with state S_l (and possibly other outputs to the environment), and will cause itself to be replaced by state S_2, which may, in some cases, involve adding new sub-systems to A.*

In some cases the causal interactions may be probabilistic rather than determinate, e.g. if part of the context that determines effects of a state consists of random noise or perturbations in lower levels of the system.

Our description is consistent with many different types of causal interactions and changes of state. Some states are representable by numerical values, whereas others involve changing structures (e.g. construction of a sentence or plan). Some causal interactions simply involve initiation, termination, excitation or inhibition, whereas others are far more complex, e.g. transmission of structured information from one sub-system to another.

So, according to VMF, each functional state S of A, such as seeing a table, waiting to eat a peach, having a particular belief, depends on enduring sub-systems of A whose states can change, as argued in Sloman (1993), and S is defined in terms of causal connections between inputs and outputs of those sub-systems in the context of various possible combinations of states of other sub-systems of A, i.e. other concurrently active functional states, as well as causal connections with S's own changing sub-states.

3. VMF and architectures

This kind of functionalist analysis of mental states is consistent with the recent tendency in AI to replace discussion of mere *algorithms* with discussion of *architectures* in which several co-existing sub-systems can interact, perhaps running different algorithms at the same time (e.g. Minsky, 1987; Brooks, 1986; Sloman, 1993). Likewise, many software engineers design, implement, and maintain virtual machines that have many concurrently active sub-systems with independently varying sub-states. A running operating system like Solaris or Linux is a virtual machine that typically has many concurrently active components. New components can be created, and old ones may die, or be duplicated. The internet is another, more complex, example. It does not appear that Block, or most philosophers who discuss functionalism, take explicit account of the possibility of virtual machine functionalism of the sort described here, even though most software engineers would find it obvious.

There are two interesting variants of VMF, restricted and unrestricted. *Restricted virtual machine functionalism* requires that every sub-state be causally connected, possibly indirectly, to inputs and outputs of the whole system A, whereas *unrestricted* VMF does not require this.[10] A philosophical view very close to restricted VMF was put forward in Ryle (1949, e.g. in the chapter on imagination), though he was widely misinterpreted as supporting a form of behaviourism.

4. Unrestricted virtual machine functionalism is biologically possible

Unrestricted VMF allows that some sub-state S or continuing sub-process is constantly altered by other sub-states that are connected with the environment even though none of the changes that occur in S affect anything that can affect the environment. An example in a computer might be some process that records statistics about events occurring in other processes, but does not transmit any of its records to any other part of the system, and no other part can access them. Unrestricted VMF even allows sub-systems that beaver away indefinitely without getting any inputs from other parts of the system. For instance, a sub-system might

[10] This is also not required by atomic state functionalism as normally conceived, since a finite state machine can, in principle, get into a catatonic state in which it merely cycles round various states forever, without producing any visible behaviour, no matter what inputs it gets.

be forever playing games of chess with itself, or forever searching for proofs of Goldbach's conjecture. Thus unrestricted VMF neither requires sensors to be capable of influencing every internal state of A, nor requires every state and process of A to be capable of affecting the external behaviour of the whole system A.

An interesting intermediate case is also allowed by VMF, namely a process that causally interacts with other processes, e.g. by sending them instructions or answers to questions, but whose internal details do not influence other processes, e.g. if conclusions of reasoning are transmitted, but none of the reasons justifying those conclusions. Then other parts of the system may know what was inferred, but be totally unable to find out why.

After a causally disconnected process starts up in A, it may be possible for new causal links to be created between it and other processes that have links to external transducers; but that is not *a precondition* for its existence. Likewise, a process that starts off with such links might later become detached (as sometimes happens to 'runaway' processes in computers).

It is often supposed that biological information processing systems must necessarily conform to *restricted* VMF, because unless some internal state or process produces external behavioural consequences under some conditions it could not possibly be selected for by evolution. However, this ignores the possibility of evolutionary changes that have side-effects apart from the effects for which they are selected, and also the possibility of evolutionary changes that have no benefits at all, but are not selected out because the environment is so rich in resources. There are some genes with positively harmful effects that continue indefinitely, such as the genes for sickle cell anaemia, which also happen to provide some protection against malaria.

We should at least consider the possibility that at a certain stage in the evolutionary history of an organism a genetic change that produces some biologically useful new information processing capability also, as a side-effect, creates a portion of a virtual machine that either runs without being influenced by or influencing other behaviourally relevant processes, or more plausibly is influenced by other states and processes but has no externally observable effects, except barely measurable increases in energy consumption. Not only is this possible, it might even be biologically useful to later generations, for instance if a later genetic change combines this mechanism with others in order to produce biologically useful behaviours where the old mechanism is linked with new ones that produce useful external behaviour. Various intermediate cases are also possible, for instance complex internal processes that produce no external effect most of the time but occasionally interact with other processes to produce useful external effects. Some processes of idle thinking, without memory records, might be like that. Another intermediate case is a process that receives and transmits information but includes more complex information processing than can be deduced from any of its input-output mappings. Completely disconnected processes might be called 'causally epiphenomenal' the largely disconnected ones and 'partly epiphenomenal'.

An interesting special case is a sub-system in which some virtual machine processes monitor, categorise, and evaluate other processes within A. This internal self-observation process might have no causal links to external motors, so that its information cannot be externally reported. If it also modifies the processes it observes, like the meta-management sub-systems described later, then it may have external effects. However it could be the case that the internal monitoring states are too complex and change too rapidly to be fully reflected in any externally detectable behaviour: a bandwidth limitation. For such a system experience might be partly ineffable.

5. Detecting disconnected virtual machine states

If everything is running on a multi-processing computer with a single central processing unit (CPU), then detached processes will have some externally visible effect because they consume energy and slow down external responses, though in some cases the effects could be minute, and even if detected, may not provide information about what the detached process is doing. If a detached process D slows others down, D is not *totally* causally disconnected. However this is a purely quantitative effect on speed and energy: the other processes need not be influenced by any of the structural details or the semantic content of the information processing in D. Where the whole system is implemented on large numbers of concurrently operating physical processors with some processor redundancy it may not be possible to tell even that the detached process is running if it slows nothing down, though delicate measurements might indicate an 'unknown' consumer of energy.

In principle it might be possible to infer processing details in such a detached virtual machine by examining states and processes in physical brain mechanisms or in the digital circuitry of a computer, but that would require 'decompiling' — which might be too difficult in principle, as people with experience of debugging operating systems will know. (E.g. searching for a suitable high level interpretation of observed physical traces might require more time than the history of the universe.)

There is a general point about the difficulty of inferring the nature of virtual machine processes from observations of low level physical traces, even when the processes do influence external behaviour. It could be the case that the best description of what the virtual machine is doing uses concepts that we have never dreamed of, for instance if the virtual machine is exploring a kind of mathematics or a kind of physical theory that humans will not invent for centuries to come. Something akin to this may also be a requirement for making sense of virtual machines running in brains of other animals, or brains of newborn human infants.

Despite the difficulties in testing for such decoupled virtual machine processes, a software engineer who has designed and implemented the system may know that those virtual components exist and are running, because the mechanisms used for compiling and running a program are trusted. (A very clever optimising compiler might detect and remove code whose running cannot affect output. But such optimisation can usually be turned off.)

Virtual machine functionalism as defined here permits possibilities that are inconsistent with conventional atomic state functionalism's presumption that only one (indivisible) state can exist at a time. Likewise VMF goes beyond simple dynamical systems theories which, as noted in Sloman (1993), talk about only one state and its trajectory. A dynamical system that has multiple coexisting, changing, interacting, attractors (like a computer) might be rich enough to support the VM architectures allowed by unrestricted VMF. Any conceptual framework that postulates only atomic (indivisible) global states will not do justice to the complexity either of current computing systems or of biological information-processing systems. We'll see below that it is also inconsistent with the phenomena referred to in talk of qualia.

6. Some VMs are harder to implement than others

The kinds of collections of concurrent interacting processes simultaneously satisfying large collections of conditional descriptions (most of which will be counterfactual conditionals at any time) will be far more difficult to implement than a process defined entirely by a sequential algorithm which always has a single locus of control. For example, Searle's thought experiment (Searle, 1980) in which he simulates a single algorithm in order to give the appearance of understanding Chinese is believable because we imagine him progressively going through the steps, with a finger keeping track of the current instruction.

It would be far more difficult for him to simultaneously simulate a large collection of interacting processes concerned with memory, perception, decision making, goal formation, self-monitoring, self-evaluation, etc., each running at its own (possibly varying) speed, and all involved in causal relationships that require substantial collections of counter-factual conditional statements to be true of all the simulated states and processes.

In other words, although virtual machine functionalism, like other forms of functionalism, allows VM states to be multiply realisable, the constraints on possible realisations are much stronger. For very complex virtual machines there may be relatively few ways of implementing them *properly* in our physical world, since many implementations including Searlian implementations, would not maintain all the required causal linkages and true counter-factual conditionals.

V: Evolvable Architectures

Different sorts of information processing systems are required for organisms with different bodies, with different needs, with different environments — and therefore different niches. Since these are virtual machines, their architectures cannot easily be inspected or read off brain structures. So any theory about them is necessarily at least partly conjectural. However, it seems that the vast majority of organisms have purely reactive architectures. A tiny subset also have deliberative capabilities, though a larger group have reactive precursors to deliberation, namely the ability to have two or more options for action activated simultaneously with some selection or arbitration mechanism to select one ('proto- deliberation'). An even smaller subset of animals seem to have meta-management capabilities (described below). These different architectural components support different varieties of what people seem to mean by 'consciousness'. Moreover, as we have seen, VMF even allows processes that satisfy the intuition that some mental states (and qualia?) have no necessary connection with perception or action.

1. Reactive architectures

Reactive architectures are the oldest type. A reactive mechanism (figure 5) is one that produces outputs or makes internal changes, perhaps triggered by its inputs and/or its internal state changes, but without doing anything that can be understood as explicitly representing and comparing alternatives, e.g. deliberating about explicitly represented future possibilities.

Many alternative reactive architectures are possible: some discrete and some continuous or mixed; some with and some without internal state changes; some with and some without adaptation or learning (e.g. weight changes in neural nets); some sequential and some with multiple concurrent processes; some with global 'alarm' mechanisms (figure 7), and some without. Some reactions produce external behaviour, while others merely produce internal changes. Internal reactions may form loops. Teleo-reactive systems (Nilsson, 1994) can execute stored plans. An adaptive system with only reactive mechanisms can be a very successful biological machine.

Some purely reactive species have a social architecture enabling large numbers of purely reactive individuals to give the appearance of considerable intelligence (e.g. termites building 'cathedrals'). The main feature of reactive systems

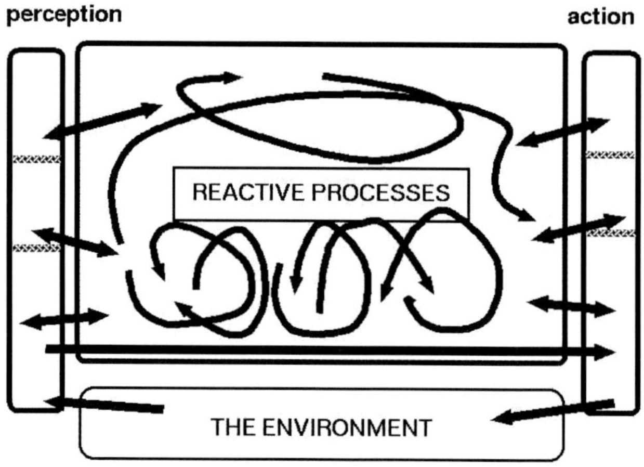

Figure 5. A simple, insect-like architecture. Arrows indicate direction of information flow. Some reactions produce internal changes that can trigger or modulate further changes. Perceptual and action mechanisms may operate at different levels of abstraction, using the same sensors and motors.

is that they lack the core ability of deliberative systems (explained below); namely, the ability to represent and reason about non-existent or unperceived phenomena (e.g., future possible actions or hidden objects). However, we have yet to explore fully the space of intermediate designs (Scheutz & Sloman, 2001).

In principle a reactive system can produce any external behaviour that more sophisticated systems can produce. However, to do so in practice it might require a larger memory for pre-stored reactive behaviours than could fit into the whole universe. Moreover, the evolutionary history of any species is necessarily limited, and reactive systems can use only strategies previously selected by evolution (or by a design process in the case of artificial reactive systems). A trainable reactive individual might be given new strategies, but it could not produce and evaluate a novel strategy *in advance* of ever acting on it, as deliberative systems with planning capabilities can. Likewise a reactive system could not discover the undesirability of a fatal strategy without acting on it. As Craik (1943) and others have noted, this limitation can be overcome by deliberative capabilities, but evolution got there first.

2. Consciousness in reactive systems

What about 'consciousness' in reactive organisms? Is a fly conscious of the hand swooping down to kill it? Insects perceive things in their environment, and behave accordingly. However, it is not clear whether their perceptual mechanisms produce information states between perception and action usable in different ways in combination with different sorts of information, as required for human-like consciousness. Purely reactive systems do not use information with the same type of flexibility as deliberative systems, which can consider non-existent possibilities. They also lack the architectural sophistication required for self-awareness, self-categorising abilities. A fly that sees an approaching hand probably does not know that it sees — it lacks meta-management mechanisms, described later. So the variety of conscious awareness that a fly has is very different from the kinds of awareness we have by virtue of our abilities to recombine and process sensory information, our deliberative capabilities, and our capacity for reflection. This more elaborate answer, rather than a simple 'yes' or 'no', is the best reply to the question 'is a fly conscious?'

In a reactive system, sensory inputs normally directly drive action-control signals, though possibly after transformations which may reduce dimensionality, as in simple feed-forward neural nets. There are exceptions: e.g. bees get information which can be used either to control their own behaviour or to generate 'messages' later on that influence the behaviour of others. We could define that as a special kind of consciousness.

3. Pressures for deliberative mechanisms

Sometimes planning is useful; in such cases, an architecture such as that depicted in figure 6, containing mechanisms for exploring hypothetical possibilities, as postulated by Craik, is advantageous. This could result from an evolutionary step

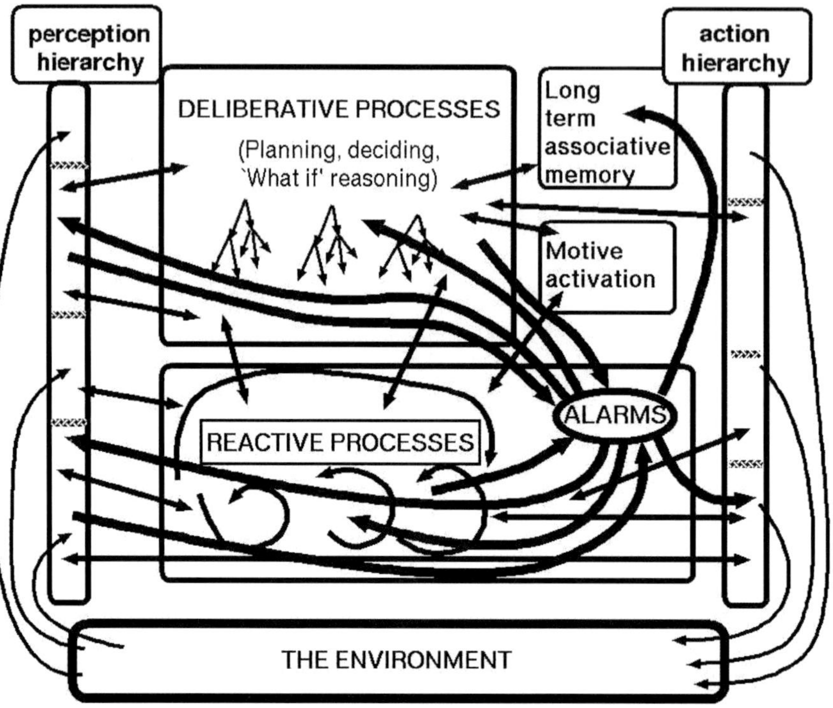

Figure 6. Reactive-deliberative architecture with 'multi-window' perception and action. Higher level perceptual and motor systems (e.g. parsers, command-interpreters) may have 'direct' connections with higher level central mechanisms. 'Alarm' mechanisms may be needed that can rapidly override and redirect slow central processes.

in which some reactive components are first duplicated then later given new functions (Maynard Smith & Szathmáry, 1999).

Deliberative mechanisms include the ability to represent possibilities (e.g. possible actions, possible explanations for what is perceived) in some explicit form, enabling alternative possibilities to be compared and one selected. Purely deliberative architectures were employed in many traditional AI systems including Winograd's SHRDLU (Winograd, 1972). Other examples include theorem provers, planners, programs for playing board games, natural language systems, and expert systems of various sorts. In robots moving at speed, deliberative mechanisms do not suffice: they need to be combined with reactive mechanisms, e.g. for dealing with sudden dangers, such as 'alarm' mechanisms.

Deliberative mechanisms can differ in various ways, e.g.:

- the forms of representations used (e.g. logical, pictorial, activation vectors — some with and some without compositional semantics);
- whether they use external representations, as in trail-blazing or keeping a diary;
- the algorithms/mechanisms available for manipulating representations;
- the number of possibilities that can be represented simultaneously (working memory capacity);
- the depth of 'look-ahead' in planning;

- the syntactic depth of descriptions and plans;
- the ability to represent future, past, concealed, or remote present objects or events;
- the ability to represent possible actions of other agents;
- the ability to represent mental states of others (linked to meta-management, below);
- the ability to represent abstract entities (numbers, rules, proofs);
- the ability to learn, in various ways;
- the variety of perceptual mechanisms (see below).

Various forms of learning are possible. Some deliberative systems can learn, and use, new abstract associations, e.g. between situations and possible actions, and between actions and possible effects. In a hybrid reactive-deliberative architecture, the deliberative part may be unable to act directly on the reactive part, but may be able to *train* it through repeated performances.

The kinds of information processing available in deliberative mechanisms can be used to define kinds of consciousness which merely reactive systems cannot have, including, for instance, 'awareness of what can happen', 'awareness of danger avoided'. The perception of possibilities and constraints on possibilities (affordances [Gibson, 1986]) is something that has not yet been adequately characterised or explained (Sloman 1989; 1996). Hybrid reactive-deliberative systems can have more varieties of consciousness, though the kinds in different parts of the architecture need not be integrated (as shown by some kinds of brain damage in humans.)

4. Pressures for multi-window perception and action

Deliberative capabilities provide the opportunity for abstract perceptual and action mechanisms that facilitate deliberation and action to evolve. New *levels of perceptual abstraction* (e.g. perceiving object types, abstract affordances), and support for *high-level motor commands* (e.g. 'walk to tree', 'grasp berry') might evolve to meet deliberative needs — hence the perception and action towers in figure 6. If multiple levels and types of perceptual processing go on in parallel, we can talk about 'multi-window perception', as opposed to 'peephole' perception. Likewise, in an architecture there can be 'multi-window action' or merely 'peephole action'. Later we'll extend this idea in connection with the third, meta-management, layer. Few current AI architectures include such multi-window mechanisms, though future machines will need them in order to have human-like consciousness of more abstract aspects of the environment.

5. Pressures for self-knowledge, self-evaluation and self-control

A deliberative system can easily get stuck in loops or repeat the same unsuccessful attempt to solve a sub-problem (one of the causes of stupidity in early symbolic AI programs with sophisticated reasoning mechanisms). One way to prevent this is to have a parallel sub-system monitoring and evaluating the

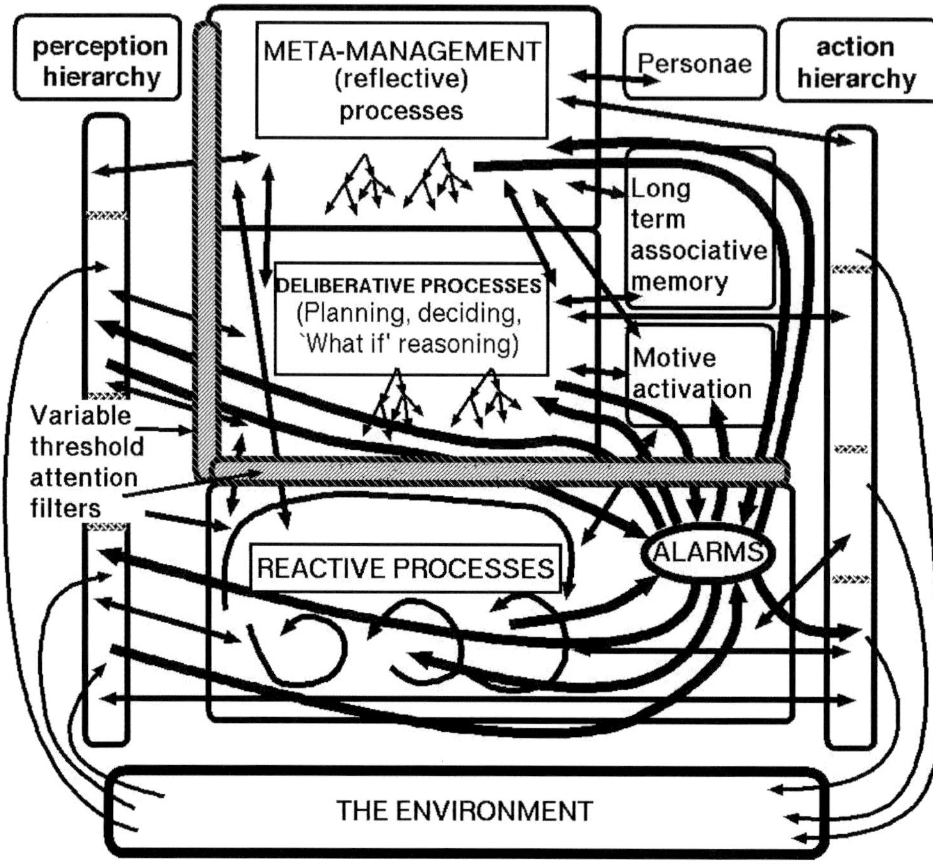

Figure 7. The H-CogAff architecture. The additional central layer supports yet more layers in perception and action hierarchies. Not all possible links between boxes are shown. Meta-management may be able to inspect intermediate states in perceptual layers.

deliberative processes. If it detects something bad happening, then it may be able to interrupt and re-direct the processing. We call this *meta-management* following Beaudoin (1994). (Compare Minsky on 'B brains' and 'C brains' in Minsky, 1987.) It is sometimes called 'reflection' by others though with slightly different connotations. It seems to be rare in biological organisms and probably evolved very late. This could have resulted from duplication and then diversification of alarm mechanisms, depicted crudely in figure 7. As with deliberative and reactive mechanisms, there are many varieties of meta-management. An interesting early example in AI is described in Sussman (1975). Psychological research on 'executive functions' (e.g. Barkley, 1997) presupposes something like meta-management, often not clearly distinguished from deliberation.

Self monitoring can include categorisation, evaluation, and (partial) control of internal processes, not just measurement. The richest versions of this evolved very recently, and may be restricted to humans, though there are certain kinds of self-awareness in other primates (Hauser, 2001).

Absence of meta-management can lead to 'stupid' reasoning and decision making both in AI systems, and in brain-damaged or immature humans, though this may sometimes be mis-diagnosed as due to lack of emotional mechanisms, as in Damasio (1994). Both the weaknesses of early AI programs with powerful deliberative capabilities and some effects of brain damage in humans that leave 'intelligence' as measured in IQ tests intact, indicate the need for a distinction between deliberative and meta-management mechanisms. Both will be needed in machines with human-like consciousness.

6. Access to intermediate perceptual data

In addition to monitoring of central problem-solving and planning processes there could be monitoring of intermediate stages in perceptual processes or action processes, requiring additional arrows going from within the perception and action towers to the top layer in figure 7. Examples would be the ability to attend to fine details of one's perceptual experience instead of only noticing things perceived in the environment; and the ability to attend to fine details of actions one is performing, such as using proprioceptive information to attend to when exactly one bends or straightens one's knees while walking. The former ability is useful in learning to draw pictures of things, and latter helps the development of various motor skills, for instance noticing which ways of performing actions tend to be stressful and therefore avoiding them — a problem familiar to many athletes and musicians. All of these processes, consistent with VFM, would need to be replicated in a machine with human-like consciousness.

7. Yet more perceptual and motor 'windows'

We conjecture that, as indicated in figure 7, central meta-management led to opportunities for evolution of additional layers in 'multi-window' perceptual and action systems: e.g. social perception (seeing someone as sad or happy or puzzled), and stylised social action (e.g. courtly bows, social modulation of speech production). This would be analogous to genetically (and developmentally) determined architectural mechanisms for multi-level perception of speech, with dedicated mechanisms for phonological, morphological, syntactic and semantic processing.

In particular, the categories that an agent's meta-management system finds useful for describing and evaluating its own mental states might also be useful when applied to others. (The reverse can also occur.) In summary:

> **Other knowledge from self-knowledge:** The representational capabilities that evolved for dealing with self-categorisation can also be used for other-categorisation, and vice-versa. Perceptual mechanisms may have evolved recently to use these representational capabilities in percepts. Example: seeing someone else as happy, or angry, seeing the duck-rabbit in figure 1 as looking left or right.

Additional requirements for coping with a fast moving environment and multiple motives (Beaudoin, 1994) and for fitting into a society of cognitively capable agents, provide evolutionary pressure for further complexity in the architecture, e.g.:

- 'interrupt filters' for resource-limited attention mechanisms,
- more or less global 'alarm mechanisms' for dealing with important and urgent problems and opportunities, when there is no time to deliberate about them,
- a socially influenced store of personalities/personae, i.e. modes of control in the higher levels of the system.

These are indicated in the figure 7, with extended (multi-window) layers of perception and action, along with global alarm mechanisms. Like all the architectures discussed so far, this conjectured architecture, (which we call 'H-CogAff', for '**h**uman-like architecture for **cog**nition and **aff**ect) could be realised in robots (in the distant future).

8. Other minds and 'philosophical' genes

If we are correct about later evolutionary developments providing high level conceptual, perceptual and meta-management mechanisms that are used both for *self*-categorisation and *other*-categorisation (using 'multi-window perception' as in the duck-rabbit picture, and in perception of attentiveness, puzzlement, joy, surprise, etc., in others), then instead of a new-born infant having to work out by some philosophical process of inductive or analogical reasoning or theory construction that there are individuals with minds in the environment, it may be provided genetically with mechanisms designed to use mental concepts in perceiving and thinking about others. This can be useful for predator species, prey species and organisms that interact socially. Such mechanisms, like the innate mechanisms for perceiving and reasoning about the physical environment, might have a 'boot-strapping' component that fills in details during individual development (see section VIII.2).

Insofar as those mental concepts, like the self-categorising concepts, refer to types of *internal* states and processes, defined in terms of aspects of the architecture that produce them, they will be architecture-based, in the sense defined in section VII, even though the implicitly presupposed architectures are probably much simpler than the actual virtual machine architectures. In other words, even if some animals (including humans!) with meta-management naturally use architecture-based concepts they are likely to be over-simplified concepts in part because they presuppose over-simplified architectures.

Nevertheless, evolution apparently solved the 'other minds problem' before anyone formulated it, both by providing built-in apparatus for conceptualising mental states in others at least within intelligent prey species, predator species and social species, and also by 'justifying' the choice through the process of natural selection, which tends to produce good designs. (Later we'll describe concepts used only to refer to the system's own internal states, in discussing qualia.)

VI: Some Implications

We have specified in outline an architectural framework in which to place all these diverse capabilities, as part of the task of exploring design space. Later we can modify the framework as we discover limitations and possible developments both for the purposes of engineering design and for explanation of empirical phenomena. The framework should simultaneously help us understand the evolutionary process and the results of evolution.

Within this framework we can explain (or predict) many phenomena, some of which are part of everyday experience and some which have been discovered by scientists:

- Several varieties of *emotions*: at least three distinct types related to the three architectural layers, namely, *primary* (exclusively reactive, such as anger), *secondary* (partly deliberative, such as frustration) and *tertiary* emotions (including disruption of meta-management, as in grief or jealousy). Some of these may be shared with other animals, some unique to humans (Wright *et al.*, 1996; Sloman & Logan, 2000; Sloman, 2001a).
- Different visual pathways, since there are many routes for visual information to be used (Sloman 1989; 1993; Goodale & Milner, 1992). The common claim that some provide 'what' information and other pathways 'where' information, may turn out to be misleading if the former provides information able to be used in deliberation (and communication) and the other provides information for control of reactive behaviours, e.g. reactive feedback loops controlling grasping or posture. There are probably many more visual pathways to be investigated.
- The phenomena discussed by psychologists in connection with 'executive functions', including frontal-lobe damage and attention disorders (e.g. Damasio, 1994; Barkley, 1997) — meta-management capabilities can be impaired while other things, including deliberative capabilities are intact.
- Many varieties of learning and development. For example, 'skill compilation' when repeated actions at deliberative levels train reactive systems to produce fast fluent actions, and action sequences. This requires spare capacity in reactive mechanisms. Information also flows in the reverse direction as new deliberative knowledge is derived from observation of one's own reactive behaviours, like a violinist discovering that changing the elevation of the elbow of the bowing arm is useful for switching between violin strings and changing the angle of the elbow moves the bow on one string.
- Limitations of self-knowledge. The model does not entail that self monitoring is perfect: so elaborations of the model might be used to predict ways in which self-awareness is incomplete or inaccurate. A familiar example is

our inability to see our own visual blind-spots. There may be many forms of self-delusion, including incorrect introspection of what we do or do not know, of how we do things (e.g. how we understand sentences), of what processes influence our decisions, and of when things happen. Experiments on change-blindness (O'Regan *et al.*, 1999), like many other psychological experiments, assume that people can report what they see. From our point of view they may be able to report only on the seeing processes as they are monitored by the meta-management system. But that need not be an accurate account of *everything* that is seen, e.g. in parts of the reactive layer. This could be the explanation of 'blindsight' (Weiskrantz, 1997): damage to some meta-management access routes prevents self-knowledge about intact (e.g. reactive) visual processes.

- The nature of abstract reasoning. The distinctions provided by the architecture allow us to make a conjecture that can be investigated empirically: mathematical development depends on development of meta-management — the ability to attend to and reflect on thought processes and their structure, e.g. noticing features of your own counting operations, or features of your visual processes.
- Evolvability: further work may help us understand some of the evolutionary trade-offs in developing these systems. (Deliberative and meta-management mechanisms can be very expensive, and require a food pyramid to support them).
- The discovery by philosophers of sensory 'qualia' (see section VIII.1).

VII: Multiple Experiencers: The CogAff Architecture Schema

The multi-disciplinary view of the whole architecture of an organism or system, and the different capabilities, states, processes, causal interactions, made possible by the various components, may lead to a particular model of human information processing. But there are different architectures, with very different information processing capabilities, supporting different states and processes. So we can expect many varieties of mentality, not just one.

Thus, we consider families of architecture-based mental concepts. For each architecture we can specify a family of concepts applicable to states, processes and capabilities supported by the architecture. The use of architecture-based concepts requires an explicit or implicit theory of the architecture supporting the states, processes and capabilities. Just as theories of the architecture of matter refined and extended our concepts of kinds of stuff (periodic table of elements, and varieties of chemical compounds) and of physical and chemical processes, so can architecture-based mental concepts, as explained in Sloman (2002), extend and refine our semantically indeterminate pre-theoretical concepts, leading to clearer concepts related to the mechanisms that can produce different sorts of mental states and processes. Note that the presupposed architectural theory need not be correct or complete. As a theory about an organism's architecture is refined and extended, the architecture-based concepts relying on that theory can also be extended.

This changes the nature of much of philosophy of mind. Instead of seeking to find 'correct' conceptual analyses of familiar mental concepts that are inherently indeterminate, such as 'consciousness', and 'emotion' we explore a space of more determinate concepts and investigate ways in which our pre-theoretical concepts related to various subsets (as the pre-theoretical concept of 'water' relates to the architecture-based concepts of H_2O and D_2O [deuterium oxide], and old concepts of chemical element relate to newer architecture-based concepts of different isotopes of an element).

New questions then supplant old ones; we can expect to replace old unanswerable questions ('Is a fly conscious?' or 'Can a foetus feel pain?') with new empirically tractable questions (e.g. 'Which of the 57 varieties of consciousness does a fly have, if any?' and 'Which types of pain can occur in an unborn foetus aged *N* months and in which sense of "being aware" can it be aware of them, if any?').

1. Towards an architecture schema

We have proposed the CogAff schema (Sloman & Logan, 2000; Sloman, 2001b) as a framework for thinking about a wide variety of information processing architectures, including both naturally occurring and artificial ones.

There are two familiar kinds of coarse divisions within components of information processing architectures: one captured in, e.g., Nilsson's 'triple tower' model (Nilsson, 1998), and the other in processing 'layers' (e.g. reactive, deliberative and meta-management layers). These orthogonal functional divisions can be combined in a grid, as indicated in figure 8. In such a grid, boxes indicate

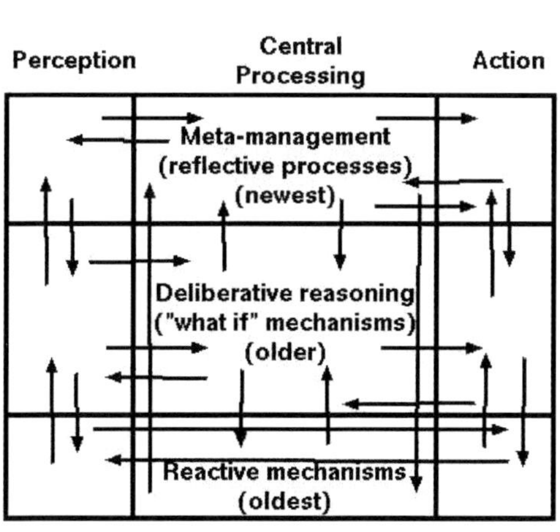

Figure 8. The CogAff schema: superimposing towers and layers.

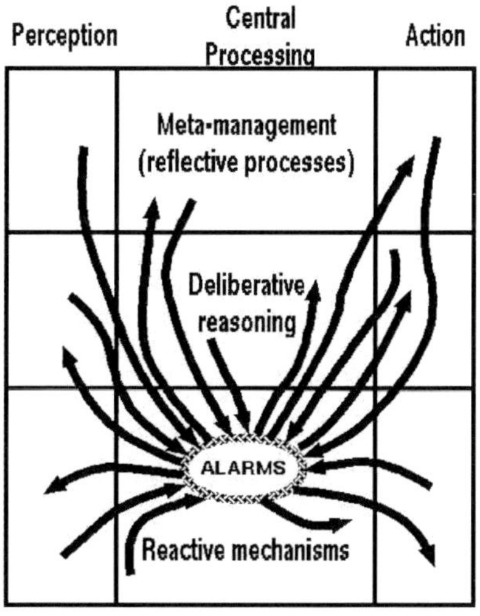

Figure 9. Elaborating the CogAff schema to include reactive alarms.

possible functional roles for mechanisms. Only a subset of all possible information flow routes are shown; cycles are possible within boxes, but not shown.

We call this superimposition of the tower and layer views the *CogAff architecture schema*, or 'CogAff' for short. Unlike H-CogAff (see section V.5), CogAff is a schema not an architecture: it is a sort of 'grammar' for architectures. The CogAff schema can be extended in various ways: e.g. see figure 9, showing alarm mechanisms added. These deal with the need for rapid reactions using fast pattern recognition based on information from many sources, internal and external, triggering wide-spread reorganisation of processing. An alarm mechanism is likely to be fast and stupid, i.e. error-prone, though it may be trainable. Such mechanisms seem to be involved in several varieties of emotions.

Different organisms, different artificial systems, may have different components of the schema, different components in the boxes, and/or different connections between components. For example, some animals, and some robots, have only the reactive layer (e.g. insects and microbes). The reactive layer can include mechanisms of varying degrees and types of sophistication — some analogue, some digital — with varying amounts of concurrency. The other two layers can also differ between species.

The schema is intended to illustrate our methodology, rather than to be a final and definitive framework for architectures. It may turn out that there are better ways of dividing up levels of functionality, or that more sub-divisions should be made — e.g. between analog and discrete reactive mechanisms, between reactive mechanisms with and without chained internal responses, between deliberative mechanisms with and without various kinds of learning, or with various kinds of formalisms; and between many sorts of specialised 'alarm' mechanisms.

2. *CogAff and varieties of consciousness*

Different architectures subsumed by the CogAff schema support different kinds of mental processes connected more or less closely with our normal notion of 'consciousness'. For example, all support some form of 'sentience', i.e. awareness of something in the environment, including the fly's awareness of your hand swooping down to catch it. If two perceptual pathways are affected when the fly detects motion of the hand, e.g. one relatively slow normal behavioural control pathway, and a rapid reaction pathway involving an 'alarm' mechanism, then the fly has two sorts of awareness of the hand. But that does not imply that it is aware of that awareness. Compare LeDoux (1996) on emotions in rats.

As mentioned earlier, architectural change can occur not only over evolutionary timescales, but also within an individual. Learning can introduce new architectural components, e.g. the ability to read music, the ability to write programs. Development of skill (speed and fluency) through practice can introduce new connections between modules, e.g. links from higher-level perceptual layers to specialist reactive modules, for instance in learning to read fluently, or developing sophisticated athletic skills. Highly trained skills can introduce new 'layer-crossing' pathways. For example, in vision, recognition of a category originally developed for deliberation can, after training, trigger fast reactions. So the architectural approach can accommodate the intuitive claim that the varieties of consciousness that are possible within an individual can develop over time. The same would happen in human-like machines.

3. Some sub-species of the CogAff schema

The schema supports many varieties of architectures that vary according to which of the 'boxes' contain mechanisms, what the mechanisms do and how they are connected.

An example sub-category of the CogAff schema is what we call 'Omega architectures'. Here only some of the possible routes through the system are used, forming roughly the shape of an Omega Ω, made up of a pipeline, with 'peephole' perception and action, as opposed to 'multi-window' perception and action (see section V.4). For examples see Albus (1981) and Cooper & Shallice (2000).

Another sub-species of CogAff is the subsumption architecture, containing several layers all within the reactive layer (Brooks, 1991). This sort of architecture is useful for understanding or designing certain relatively primitive sorts of organisms (e.g. insects, fish, crabs?) and robots.

By locating various architectures within a common schematic framework, we facilitate the task of comparing and contrasting the various forms of consciousness which the architectures support. Comparing designs and analysing trade-offs is a better way to proceed than arguing endlessly about which architecture is 'correct', or 'sufficient for consciousness'. Instead we replace our single cluster concept with a variety of more refined, architecture-based concepts about which more productive empirical investigations are possible. Of course, conceptual indeterminacy may never be eliminated completely, because of the possibility of unanticipated subdivisions with surprising combinations of features.

VIII: Some Objections

1. An architecture-based explanation of qualia?

At this point, some readers will be wondering how these architectural notions make any headway on the infamous problems of consciousness, e.g. the problem of qualia - the private, ineffable way things seem to us. We have planted pointers to our answer at various points in preceding sections. In section VI, we suggested that an architecture-based explanation of qualia is possible. Primarily we explain qualia by providing an explanation of *the phenomena that generate philosophical thinking of the sort found in discussions of qualia*. Note that we are not talking merely about explaining behaviour, for we have repeatedly discussed explanations of how internal, possibly externally undetectable, states and processes can occur in certain virtual machine architectures.

The concept of 'qualia' arose out of philosophical discussions of our ability to attend to aspects of internal information processing (internal self-awareness). That possibility is inherent in any system that has the H-CogAff architecture (see

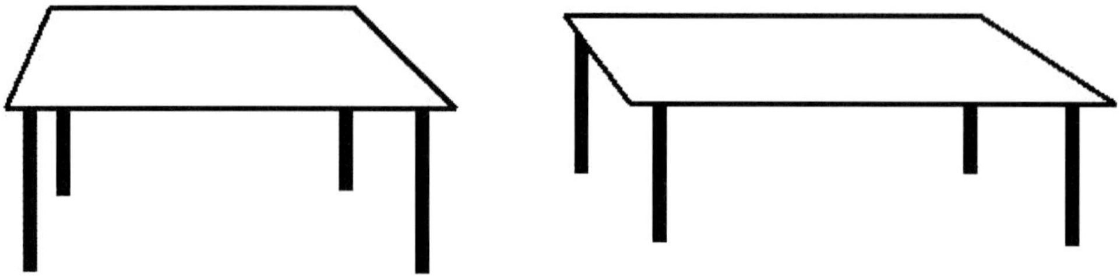

Figure 10. Noticing two perspectives on the same object is one route to concerns about qualia.

section V.5), though different varieties of the phenomenon will be present in different architectures, depending on the forms of representation and modes of monitoring available to meta-management. Some forms will provide the ability to attend not only to what is perceived in the environment, but to also features of the *mode of perception* that are closely related to properties of intermediate sensory data-structures, as indicated by some of the arrows in figure 7 from the perceptual tower to the centre.

Consider perceiving a table. Most adults (though not young children) can attend not only to the table and its fixed 3-D shape, but also to the 2-D *appearance* of the table in which angles and relative lengths of lines change as you change your viewpoint (or the table is rotated; see figure 10). The appearance can also change as you squint, tap your eyeball, put on coloured spectacles, etc. This is exactly the sort of thing that led philosophers (and others) to think about qualia (previously called 'sense data') as something internal, non-physical, knowable only from inside, etc. If meta-management processes have access to intermediate perceptual states, then this can produce self-monitoring of sensory contents, leading robot philosophers with this architecture to discover 'the problem(s) of qualia'. And the same would go for *anything* which has that architecture: six robots with the H-CogAff architecture discussing various aspects of their experience of the same table seen from different viewpoints could get bogged down discussing consciousness, just like six blind philosophers.

> **What qualia are:** qualia are what humans or future human-like robots refer to when referring to the objects of internal self observation.

2. Architecture-based and architecture-driven concepts

We have used 'architecture-based' to describe the concepts used by us to refer ('from outside') to possible states and processes in an information processing machine, where our concepts are defined in terms of what components of the architecture can do. An individual can also use such concepts to refer to its own states, as suggested earlier. But there is another way in which an information processing system can refer to its own states, which explain some aspects of notions of qualia.

Concept formation is a huge topic, but, for now, consider the likelihood that in many organisms there are processes of concept formation which emerge from interactions between a self-organising classification system and the information fed into it. A well known example of a mechanism that can achieve this is a Kohonen net (Kohonen, 1989). We describe as 'architecture-driven' sets of concepts created within an architecture as part of the individual history of the organism or machine. If individual A1 develops its own concepts used to describe internal states of another agent A2 on the basis of assumptions about the information processing architecture of A2, then the concepts are architecture-driven in relation to A1, and architecture-based in relation to A2, or any other system with the assumed architecture. It is possible for A1 to use architecture-driven architecture-based concepts to refer to itself. Architecture-driven concepts can refer to many different sorts of things, e.g. colours and shapes of objects in the environment, tastes, tactile qualities of objects, etc., if the concepts are developed by an organism on the basis of perceptual experience of those objects.

3. The privacy and ineffability of qualia

Now suppose that an agent A with a meta-management system, as in figure 7, uses a self-organising process to develop concepts for categorising its own internal virtual machine states as sensed by internal monitors. These will be architecture-driven concepts, but need not be architecture-based if the classification mechanism does not use an implicit or explicit theory of the architecture of the system it is monitoring, but merely develops a way of organising its 'sensory' input data. If such a concept C is applied by A to one of its internal states, then the only way C can have meaning for A is in relation to the set of concepts of which it is a member, which in turn derives only from the history of the self-organising process in A. These concepts have what Campbell (1994) refers to as 'causal indexicality'. This can be contrasted with what happens when A interacts with other agents in such a way as to develop a common language for referring to features of external objects. Thus A could use 'red' either as expressing a private, causally indexical, concept referring to features of A's own virtual-machine states, or as expressing a shared concept referring to a visible property of the surfaces of objects.

This means that if two agents A and B have each developed concepts in this way, then if A uses its causally indexical concept Ca, to think the thought 'I am having experience Ca', and B uses its causally indexical concept Cb, to think the thought 'I am having experience Cb' the two thoughts are intrinsically private and incommunicable, even if A and B actually have exactly the same architecture and have had identical histories leading to the formation of structurally identical sets of concepts. A can wonder: 'Does B have an experience described by a concept related to B as my concept Ca is related to me?' But A cannot wonder 'Does B have experiences of type Ca', for it makes no sense for the concept Ca to be applied outside the context for which it was developed, namely one in which A's internal sensors classify internal states. They cannot classify states of B.

When different agents use architecture-driven causally indexical concepts that are produced by self organising classifiers, to classify *internal states of a virtual machine*, and which are not even partly explicitly defined in relation to some underlying causes (e.g. external objects or a presumed architecture producing the sensed states), then there is nothing to give those concepts any user-independent content, in the way that our colour words have user-independent content because they refer to properties of physical objects in a common environment. Thus self-referential architecture- driven concepts used by different individuals are strictly non-comparable: not only can you not know whether your concepts are the same as mine, the question is *incoherent*. If we use the word 'qualia' to refer to the virtual machine states or entities to which these concepts are applied, then asking whether the qualia in two experiencers are the same would then would be analogous to asking whether two spatial locations in different frames of reference are the same, when the frames are moving relative to each other. But it is hard to convince some people that this makes no sense, because the question is grammatically well-formed. Sometimes real nonsense is not *obvious* nonsense.

We have now indicated how the process of coming to think about and ask questions about qualia is explained by the nature of the architecture of the thinker. The process arises when architecture-driven causally indexical concepts produced by a self-monitoring sub-system refer internally. In talking about these concepts we are using architecture-based concepts.

Is this a new kind of explanation? Perhaps, although it seems to have some similarities with one of Hume's explanations of the concept of 'causation'. Hume attempted to explain what a cause was by looking at what it was *about us* that made us look at the world in terms of that concept. Kant attempted to reformulate this by arguing that having a concept of causation was *necessary* for having experiences of an objective external world.

Becoming interested in and puzzled by qualia is a product of sophisticated biological mechanisms, whose nature explains why some questions about qualia can have no answers. We do not say, as some positivist philosophers would, that all talk about qualia is nonsensical: rather we try to show what can and cannot be sensibly said about them.

Some robots with our information processing architecture will discover qualia and be puzzled about them. The more intelligent ones should accept our explanation of how that happens.

4. Is something missing?

There will always be people who are convinced that this sort of project inevitably fails to answer the questions about consciousness which *they* think are the real ones. Often these are people who say of consciousness some of the things we listed in table 1, e.g.

- It's indefinable, knowable only through having it;
- It's what it is like to be something (hungry, in pain, happy, a bat . . .);
- It's possibly absent in something (behaviourally, functionally) indistinguishable from us (zombies).

The fact that many people do think like this is *part of what needs to be explained* by any adequate theory of consciousness. Our explanation is that it is a side-effect of some of the processes made possible by the existence of a meta-management layer which allows an information processing system to attend to aspects of its own internal functioning, e.g. some of the intermediate states in its sensory mechanisms, and use architecture-driven concepts to describe them. Therefore, we offer yet another conjecture:

> **The inevitability of consciousness talk:** When robots have suitably rich internal information processing architectures some of them will also feel inclined to talk about consciousness, and qualia, in a way similar to the way we do.

This isn't a particularly new idea; science fiction writers thought of this long ago. It implies that even if philosophical theories about qualia, about 'what it is like to be something', involve much confusion and even error, it does not follow that they are completely wrong. They are based on a correct *partial* view of how minds work.

5. *Zombies*

Many will be *convinced* that something is left out, and will express this conviction in the form of the 'zombie' argument: a robot could have all the information processing capabilities described here (albeit sketchily) and still lack 'this' — said attending inwardly, using human meta-management capabilities. That is, the robot might be a *zombie* (for a survey of arguments see Chalmers, 1996). Usually such arguments fail to distinguish the different forms of functionalism we have distinguished. It is true that a system could have all the *externally observable* behaviours, i.e. the same input–output mappings as humans have, without having the sort of consciousness discussed here. However the claim that a robot can have all the *internal* information processing capabilities humans have, and can have internal sub-states satisfying all the same sets of causal relationships and counterfactual conditionals, and still lack something we have, becomes incoherent if nothing specific can be said about what is missing.

When internal processing of a human-like virtual machine is described *in great detail*, including the meta-management abilities involved in thinking about qualia, including, where appropriate the 'disconnected' and non-reportable states and processes, it is not clear that anything intelligible is left over: the description of a zombie as being just like us in all its capabilities yet unlike us in experiencing qualia becomes incoherent.

6. Are we committed to 'computationalism'?

It is important to distinguish two questions:

(1) Is there any information processing virtual machine architecture that is sufficient to produce mental states and processes like ours?
(2) Which, if any, of these virtual machines can be implemented on a computer (of the sort that we currently know how to build)?

It is often assumed, wrongly, that a negative answer to 2 implies a negative answer to 1. That's because many people do not appreciate that the general notion of an information processing machine is not defined in terms of computers — computers just happen to be the best tools we currently have (and even they often include sub-systems that do not conform to the standard notion of a computer, e.g. analogue components and various transducers. In future we may invent new kinds of information processing engines, as different from computers as computers are from mechanical calculators. It might turn out that certain sorts of virtual machine architectures are adequate for the implementation of all typical adult human mental phenomena, but that no digital computer is able to support them all. We might find other kinds of previously unknown computers in biological systems.

Finding out answers requires us first to clarify meanings of the questions and the available answers. We know no better method than the method outlined here.

7. Falsifiability? Irrelevant.

What we have proposed isn't directly confirmable or falsifiable as a scientific hypothesis. Some, perhaps armed with a superficial reading of Popper (1934), might take this lack of falsifiability as grounds for rejecting our proposal. But to ask: 'Is the theory falsifiable?' is to ask the wrong question. Within the proposed framework we can make simultaneous progress in science (several different sciences) and philosophy, including investigating relationships between brain mechanisms and the virtual machine architectures described here. What's more important than (immediate) falsifiability is the ability to generate large numbers of different, non-trivial consequences about what is possible, e.g. implications about possible types of learning, about possible forms of perception, about possible types of emotions.

You can't empirically refute statements of the form 'X can happen'. But they can open up major new lines of research and unify old ones. Following Popper and Lakatos we need to ask whether this will turn out to be a *progressive* or a *degenerative* research programme. We hope we have given reason to believe that it is the former.

Acknowledgements

This work is partly funded by grant F/94/BW from the Leverhulme Trust, for research on 'Evolvable virtual information processing architectures for human-like minds'. The ideas presented here were inspired by work of many well known philosophers, and developed with the help of Matthias Scheutz and also many students, colleagues and friends, including Margaret Boden, Pat Hayes, Steve Allen, John Barnden, Luc Beaudoin, Catriona Kennedy, Brian Logan, Riccardo Poli and Ian Wright. We have also learnt much from the empirical work of colleagues in the School of Psychology at the University of Birmingham, including Glyn Humphreys, Jane Riddoch and Alan Wing. We are grateful to the managing editor, Anthony Freeman, for his help and patience.

References

Albus, J. (1981), *Brains, Behaviour and Robotics* (Peterborough: Byte Books, McGraw Hill).

Barkley, R.A. (1997), *ADHD and the Nature of Self-control* (New York: The Guildford Press).

Beaudoin, L. (1994), 'Goal processing in autonomous agents', PhD thesis, School of Computer Science, The University of Birmingham (http://www.cs.bham.ac.uk/ research/ cogaff/).

Block, N. (1995), 'On a confusion about the function of consciousness', *Behavioral and Brain Sciences*, **18**, pp. 227–47.

Block, N. (1996), 'What is functionalism?', http://www.nyu.edu/gsas/dept/philo/faculty/block/papers/functionalism.html (from *The Encyclopedia of Philosophy Supplement*, Macmillan, 1996).

Boden, M.A. (2000), 'Autopoesis and life', *Cognitive Science Quarterly*, **1** (1), pp. 115–43.

Brooks, R. (1986), 'A robust layered control system for a mobile robot', *IEEE Journal of Robotics and Automation* RA-2, pp. 14–23.

Brooks, R.A. (1991), 'Intelligence without representation', *Artificial Intelligence*, **47**, pp. 139–59.

Campbell, J. (1994), *Past, Space and Self* (Cambridge, MA: MIT Press)

Chalmers, D.J. (1996), *The Conscious Mind: In Search of a Fundamental Theory* (New York: Oxford University Press).

Chrisley, R. (1995), 'Non-conceptual content and robotics: Taking embodiment seriously', in *Android Epistemology*, ed. K. Ford, C. Glymour, P. Hayes (Cambridge, MA: AAAI/MIT Press).

Chrisley, R. (2000), 'Transparent computationalism', in *New Computationalism: Conceptus-Studien 14*, ed. M. Scheutz (Sankt Augustin: Academia Verlag).

Cohen, L.J. (1962), *The Diversity of Meaning* (London: Methuen & Co. Ltd.).

Cooper, R. and Shallice, T. (2000), 'Contention scheduling and the control of routine activities', *Cognitive Neuropsychology*, **17** (4), pp. 297–338.

Craik, K. (1943), *The Nature of Explanation* (Cambridge: Cambridge University Press).

Damasio, A. (1994), *Descartes' Error, Emotion Reason and the Human Brain* (New York: Grosset/Putnam Book).

Gibson, J. (1986), *The Ecological Approach to Visual Perception* (Hillsdale, NJ: Lawrence Erlbaum Associates; originally published in 1979).

Goodale, M. and Milner, A. (1992), 'Separate visual pathways for perception and action', *Trends in Neurosciences*, **15** (1), pp. 20–25.

Hauser, M. (2001), *Wild Minds: What Animals Really Think* (London: Penguin).

Kim, J. (1998), *Mind in a Physical World* (Cambridge, MA: MIT Press).

Kohonen, T. (1989), *Self-Organization and Associative Memory* (Berlin: Springer-Verlag).

LeDoux, J. (1996), *The Emotional Brain* (New York: Simon & Schuster).

Maynard Smith, J. and Szathmáry, E. (1999), *The Origins of Life: From the Birth of Life to the Origin of Language* (Oxford: Oxford University Press).

Minsky, M.L. (1987), *The Society of Mind* (London: William Heinemann Ltd.).

Nilsson, N. (1994), 'Teleo-reactive programs for agent control', *Journal of Artificial Intelligence Research*, 1, pp. 139–58.

Nilsson, N. (1998), *Artificial Intelligence: A New Synthesis* (San Francisco: Morgan Kaufmann).

O'Regan, J., Rensink, R. and Clark, J. (1999) , 'Change-blindness as a result of "mudsplashes"', *Nature*, **398** (6722), p. 34.

Peterson, D. (ed. 1996), *Forms of Representation: An Interdisciplinary Theme for Cognitive Science* (Exeter: Intellect Books).

Popper, K. (1934), *The Logic of Scientific Discovery* (London: Routledge).

Rose, S. (1993), *The Making of Memory* (London: Bantam Books).

Ryle, G. (1949), *The Concept of Mind* (London: Hutchinson).
Scheutz, M. (1999), 'The missing link: Implementation and realization of computations in computer and cognitive science', PhD thesis, Indiana University (University of Michigan Microfiche).
Scheutz, M. and Sloman, A. (2001), 'Affect and agent control: Experiments with simple affective states', in *Intelligent Agent Technology: Research and Development* (New Jersey: World Scientific).
Searle, J. (1980), 'Minds brains and programs', *The Behavioral and Brain Sciences*, **3**, pp. 417–57 (with commentaries and reply by Searle).
Sloman, A. (1989), 'On designing a visual system (Towards a Gibsonian computational model of vision)', *Journal of Experimental and Theoretical AI* **1** (4), pp. 289–337.
Sloman, A. (1993), 'The mind as a control system', in *Philosophy and the Cognitive Sciences*, ed. C. Hookway & D. Peterson (Cambridge: Cambridge University Press).
Sloman, A. (1996), 'Actual possibilities', in *Principles of Knowledge Representation and Reasoning: Proceedings of the Fifth International Conference (KR '96)*, ed. L. Aiello S. Shapiro (Boston, MA: Morgan Kaufmann Publishers).
Sloman, A. (2000), 'Interacting trajectories in design space and niche space: A philosopher speculates about evolution', *Parallel Problem Solving from Nature — PPSN VI, Lecture Notes in Computer Science, No 1917* (Berlin: Springer-Verlag).
Sloman, A. (2001a), 'Beyond shallow models of emotion', *Cognitive Processing: International Quarterly of Cognitive Science*, **2** (1), pp. 177–98.
Sloman, A. (2001b), 'Varieties of affect and the CogAff architecture schema', in *Proceedings Symposium on Emotion, Cognition, and Affective Computing AISB'01 Convention', York*, ed. C. Johnson, pp. 39-48.
Sloman, A. 2002, 'Architecture-based conceptions of mind', in *In the Scope of Logic, Methodology, and Philosophy of Science (Vol II)*, pp. 403–27 (Dordrecht: Kluwer; Synthese Library Vol. 316).
Sloman, A. and Logan, B. (2000), 'Evolvable architectures for human-like minds', in *Affective Minds*, ed. G. Hatano, N. Okada and H. Tanabe (Amsterdam: Elsevier).
Sloman, A. and Scheutz, M. (2001), 'Tutorial on philosophical foundations: Some key questions', in *Proceedings IJCAI-01', AAAI, Menlo Park, California*, pp. 1–133.
http://www.cs.bham.ac.uk/~axs/ijcai01.
Smith, B. (1996), *On the Origin of Objects* (Cambridge, MA: MIT Press).
Sussman, G. (1975), *A Computational Model of Skill Acquisition* (American Elsevier).
Waismann, F. (1965), *The Principles of Linquistic Philosophy* (London: Macmillan).
Weiskrantz, L. (1997), *Consciousness Lost and Found* (Oxford: Oxford University Press).
Winograd, T. (1972), 'Procedures as a representation for data in a computer program for understanding natural language', *Cognitive Psychology* (later published as a book *Understanding Natural Language*, Academic Press, 1972).
Wittgenstein, L. (1922), *Tractatus Logico-Philosophicus*, trans. C.K. Ogden (London: Routledge and Kegan Paul).
Wittgenstein, L. (1953), *Philosophical Investigations* (Oxford: Blackwell; 2nd edition 1958).
Wright, I., Sloman, A. and Beaudoin, L. (1996), 'Towards a design-based analysis of emotional episodes', *Philosophy Psychiatry and Psychology*, **3** (2), pp. 101–26.
Reprinted in *Artificial Intelligence: Critical Concepts in Cognitive Science*, Vol IV, ed. R.L.Chrisley (London: Routledge 2000).

Luc Steels

Language Re-Entrance and the 'Inner Voice'

As soon as we stop talking aloud, we seem to experience a kind of 'inner voice', a steady stream of verbal fragments expressing ongoing thoughts. What kind of information processing structures are required to explain such a phenomenon? Why would an 'inner voice' be useful? How could it have arisen? This paper explores these questions and reports briefly some computational experiments to help elucidate them.

Keywords: consciousness, origins of language, inner voice, robotic models.

I: Introduction

Although consciousness is sometimes believed to involve a single explanatory principle (and would therefore be something like gravity, see Penrose,1990), it has so far proven more fruitful to regard it as a combination of many different aspects. These include: unifying different views of reality experienced through different sensory modalities into a coherent picture, focusing attention, feeling emotions, acting coherently by combining top-down goals with bottom-up reactive behaviour, forming thoughts and expressing them in words, building and maintaining a self-model, intuiting sudden insights (the 'Aha' experience), making and justifying ethical decisions, and having a subjective phenomenal experience of the self.

The main reason for analysing consciousness this way comes from convincing evidence that there is a functional specialization of brain areas and regions. Visual awareness is correlated to coalitions of active neurons in the neocortex (Crick and Koch, 1995), speech production and perception shows strong activations of Broca's and Wernicke's area, emotion heavily exercises the amygdala and the hormonal system (LeDoux, 1998). Neuroscience has concentrated enormous efforts in recent decades in order to map these functional specializations, i.e. to find the neural correlates of various mental capacities, including those associated with consciousness (Edelman, 1989).

However knowing where some activity relevant to a certain mental function takes place is not yet a satisfactory explanation. Food is digested in the stomach and the liver. Fine. But we want to know *how* the digestion takes place, *why* food needs to be digested, and *why* the digestive products are necessary for survival. Moreover mental functions might be less localized than they are thought to be (Rosenfield, 1992), just as food digestion actually involves the whole body and its behaviour.

As Dennett has suggested (Dennett, 1991), building very detailed operational models explicating the information processing necessary and sufficient for certain mental functions is one way to address the how and why of some aspects of conscious processes, and AI researchers have indeed been doing this, sometimes with considerable success. The work on modelling (shared) attention is a good example of this approach.

Children are already capable of establishing shared attention and turn-taking after a few months of life, and it forms a necessary basis for the social learning of language and meaning (Tomasello, 1999). The many components involved in attention sharing are beginning to be understood and several AI researchers have recently built integrated robotic systems that combine a large set of these mechanisms to achieve attentional behaviour (Scassellati, 1998). They include the ability to pick from a visual image stream the objects that are salient in a particular context, such as faces, separate out sound sources and identify visually the object responsible for the sound, keep tracking objects as they move, and observe what a person is attending to by detecting pointing gestures and the orientation of the face. These mechanisms are being integrated on a new generation of humanoid robots (such as KISMET [Breazeal, 1998] or the Sony SDR robot [Kuroki *et al.*, 2001]), yielding impressive human-like performance.

Are the robots employing these mechanisms conscious? Most people would say no — even though it is quite extraordinary to interact with them. But these developments show that whatever consciousness 'really' is, some of the behaviours often associated with having consciousness can be unravelled, and their information processing foundations understood.

This paper adopts the same strategy for studying another aspect of human behaviour associated with consciousness. Every normal human being, including deaf people, report hearing almost constantly a silent voice when they are not overtly speaking. This inner voice jumps from one topic to the other. It seldom produces very well formed sentences but rather fragmented bits and pieces of thought, as captured by Joyce's *Ulysses*. The inner voice is there as soon as we wake up and particularly when we are trying to write or form new thoughts. People fluent in several languages report that their inner voice occasionally speaks another language. The inner voice is closely linked to the self, indeed we have the feeling that it is the 'I' which is talking, like in a first person narrative. It is also possible to hear the voice of somebody else, such as when we imagine a dialogue or hear people speak in a dream.

The questions raised in this paper are going to be the following: Can we unravel the information processing structures achieving this mental functioning? How exactly might such an inner voice be produced? And why would an inner voice have arisen in the course of evolution? What are additional benefits once such an inner voice is there?

In summary, my argument is the following:

(1) Achieving an inner voice is technically not so difficult. The language system must be turned onto itself. Output from the subsystem that produces speech must be fed back into the subsystem that understands it. Many psycholinguistic models going back to the nineteenth century show this feature and some of our recent computer simulations do so as well.

(2) The why question also has a straightforward answer. The re-entrant mapping of the language system is a means for bootstrapping language towards higher complexity, indeed I believe it is a prerequisite. Many psycholinguists have argued that natural language arose as an externalization of the internal language of thought (Fodor, 1975). I will argue just the opposite. Internal language is a side effect of being able to use and learn external language.

(3) The 'inner voice mechanism' is not only useful in language construction and language learning. It can also be used to come to terms with our experiences. It enables a simulation of the thought processes of others and thus provides the foundation for complex social behaviour.

The remainder of this paper discusses these points in more detail. I will report on some computational and robotic experiments, trying to avoid as much as possible over-technical aspects.

II: Language Communication

1. Language as a complex adaptive system

We need a theory of language which differs from the one common in contemporary Chomskyan linguistics. Generative grammar has been conceived as a descriptive model of the syntactic regularities of languages, whereas we need a model of how language production and interpretation works in concrete settings. There are three characteristics of language which I want to emphasize:

1. Language arose for communication. We speak primarily because we want to influence the thoughts and behaviours of somebody else (Austin, 1975). This does not mean that there are no derived uses of language, as in religious ceremonies. But denying the communicative function of language makes it incomprehensible why we speak and also why languages are the way they are. So, the generative source of speech does not lie in syntax, as in the Chomskyan model, but in subsystems that conceptualize and select meaning.

2. Natural language does not fit with the simplest form of the Shannon model of communication. In a Shannon model, there is a sender and receiver, a

message, and a message content. Production and interpretation is considered to be equal to a coding and decoding process. The content is assumed to be completely present in the message. In contrast, communication in natural language assumes an intelligent receiver, i.e. an agent that shares the same context and can fill in a large part of what the sender wants to transmit (Sperber and Wilson, 1986). This is why the computer processing of natural language is not really possible, unless a computer or robot has the same inferential powers and shares the same environment and contexts as humans.

The inferential aspect of natural language communication is its main power but also its principal weakness. It is more powerful than a Shannon system because much more complex information can be communicated with fewer means. But it is weaker because much more processing needs to be done both by the speaker (who has to guess what the hearer might already know or be able to infer) and by the listener (who has to guess what the speaker might mean). Moreover there is a higher risk of misunderstanding if speaker and hearer do not sufficiently share a common ground.

3. Language is in constant evolution. Every verbal interaction is a creative one and lexical and grammatical regularity is emergent and transient (Hopper, 1991). Speakers and hearers create new conventions for the occasion (Clark and Brennan, 1991) and these might then propagate to the rest of the community. Innovations include new words, new meanings for existing words, new grammatical constructions, and new patterns of dialogue. These micro-changes in language are perhaps not visible to language participants but they are blatantly obvious when one examines language usage even over a very short stretch of time. A person's language is always in flux and any model of language processing must include components to explain how participants in a conversation adapt the language to suit their needs.

2. The semiotic cycle

Obviously the formulation of an utterance and its interpretation involve highly complex processes, particularly when the communication is grounded in a concrete situation and therefore implies perception and action. Most of these processes are completely unconscious. We have no direct access to the intermediary structures nor to the grammatical knowledge we all possess. And it is best that way. Language processing is so complex that it would not be possible to keep every intermediary step in mind, particularly because processing must be fast enough to respond fluently in dialogue.

The main processes involved in language comprehension and production are displayed in the semiotic cycle (see figure 1). They take place against the background of turn-taking and attention sharing behaviours and are of course strongly intermingled and influencing each other.

(1) *Grounding*. The first set of processes carried out by the speaker must maintain a connection between the internal factual memory and the states and actions in the world the speaker wants to talk about. They

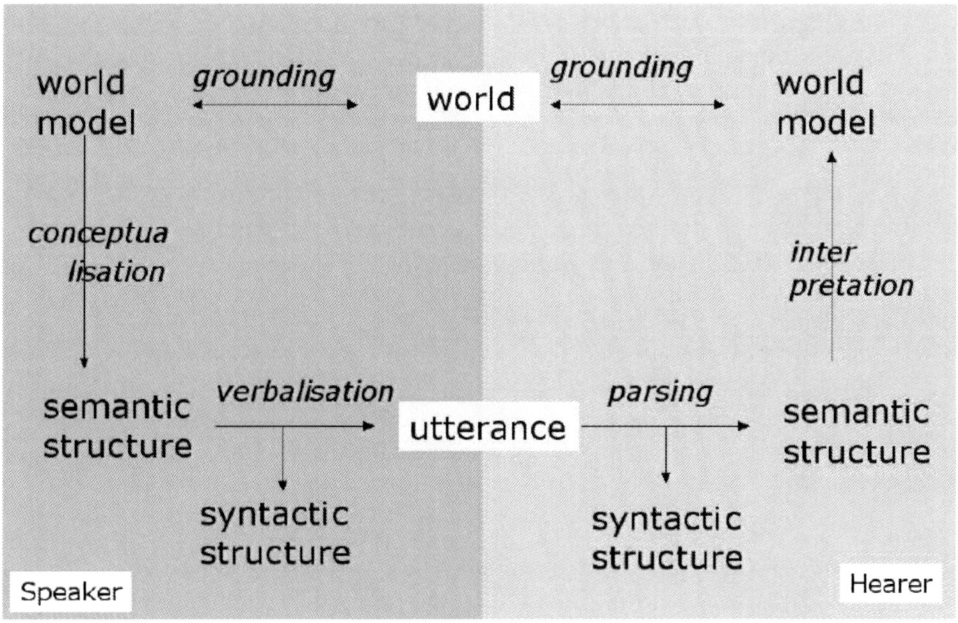

Figure 1. The semiotic cycle

Left: processes carried out by the speaker. Right: processes carried out by the hearer.

include segmentation, feature extraction, object recognition, event classification, object tracking, object manipulation, etc.

(2) *Conceptualisation*. The second set of processes must select what needs to be said and conceptualize the world in a way that it can be translated to natural language (Talmy, 2000). This requires inference and possibly additional perception or action. For example, if we say 'the car is in front of the tree', we have conceptualized the tree as having a front which is directed towards us, and the car as being in between ourselves and this front. This conceptualisation is language dependent. In some African languages, the front of the tree is in the same direction as the face of somebody looking at a tree, so 'the car in front of the tree' now means behind the tree (Heine, 1997).

(3) *Verbalisation*. This is the set of processes that take a semantic structure and render it through a series of mappings into a surface form. This is an enormously complex process, bringing to bear a vast array of lexical, grammatical, morphological, phonological and phonetic conventions, which are obviously language-dependent. The final outcome of verbalisation is transmitted as an utterance to the hearer.

(4) *Parsing*. The hearer must reconstruct a semantic structure from the utterance that is transmitted. This process is again highly elaborate due to the complexity of natural language and the presence of ambiguities. The steps in parsing cannot simply be chained linearly but must operate as a constraint propagation process.

(5) *Interpretation.* The hearer must confront the semantic structure resulting from the parsing process with his factual memory of the world and understanding of the dialogue context in order to find a correct interpretation, i.e. an interpretation of the utterance that fits with his expectations and observations.
(6) *Grounding.* The hearer must therefore also maintain a connection through perception and action between his internal factual memory and the states of the world, including the states of the speaker.

At present we can construct computational models of all aspects of this cycle (at least in a simplified form), and for our experiments on embodied communication with autonomous robots this is what we do (Steels, 2001a,b; Steels and Kaplan, 2001). The semiotic cycle is embedded in language games, in which speaker and hearer take turns trying to achieve a particular goal (such as drawing attention to an object in the environment through verbal means).

3. Inventing and learning

The semiotic cycle captures the routine application of existing language conventions, but, as argued earlier, this routine application is hardly ever possible. New situations and new meanings that have never been expressed before come up all the time and so language users are constantly expanding and learning new language. This requires two additional processes:

1. *Language invention.* It is unavoidable that situations arise where the speaker does not know an adequate well-entrenched convention for expressing certain meanings. Either the meanings have never been expressed before (think about all the operations that we can now do through Internet which would be unthinkable fifty years ago), or their conventional expression might give rise to misunderstandings in the present context. Moreover the speaker may occasionally want to avoid worn out phrases to better capture the attention of the listener. So a set of processes is needed that implements strategies to invent new language conventions on the spot. The simplest example of a language invention strategy is to chose a new word to express a predicate. Usually the new word is chosen by analogy with an existing one. For example, a computer pointing device is called a mouse by analogy with the shape of a mouse, or clicking on a hyperlink is called surfing by analogy with catching waves. Invention also happens for grammatical constructions, even though it is perhaps less visible. For example, if no satisfying expression exists to express a certain temporal relation, another word or another grammatical construction might be adopted for this purpose. This is how the auxiliary 'will' or the construction 'is going to' became gradually adopted in English for the grammatical expression of future (Traugott and Heine, 1991).

2. *Language learning.* A language listener must be prepared to learn new conventions at any moment. As argued earlier, a large part of the meaning of an utterance is implicit and must be actively reconstructed. This is only possible because he shares the same situation as well as the broader context with the speaker, makes similar inferences, and conceptualizes the world in a similar way. These

inferential capabilities make it possible for the listener to learn new bits of language. Suppose someone says (while sitting in front of a computer), 'Can you hand me the mouse', and the listener does not know what a mouse is. If the speaker then points to or grabs the mouse herself, the listener can easily infer that this is what is called a mouse. The analogy between a mouse and the computer's pointing device is very suggestive and therefore not so hard to guess or to remember afterwards.

When a population of speakers and listeners applies language invention and language learning on a large scale and in a cumulative fashion, a communication system progressively emerges. In a series of computer and robotic simulations, we have been experimenting with such processes and consistently seen this phenomenon (Steels *et al.*, 2002). The artificial agents effectively negotiate language conventions. They monitor whether a particular meaning-form association has been successful in communication by keeping a score. Agents should prefer the meaning-form associations that had the highest score in the past because this gives the best chance for success in the future. This leads to the spontaneous self-organization of a shared set of conventions due a positive feedback between use and success: The score of a convention which contributed to successful communication increases for the agents that use it, hence they will use this convention even more. Agents that did not know the convention will hear it and eventually all agents will have adopted the same convention.

We have done a number of experiments to simulate both the processes in the semiotic cycle and the invention and learning processes outlined in this subsection. These are part of a more encompassing research programme trying to build computational and robotic models for the emergence of grammar. In our experiments, two robotic heads watch scenes, such as the one shown in figure 2 (from Steels and Baillie, 2002) in which a hand picks up a red object. Then they play a language game. For example, a guessing game in which the speaker chooses an object in the scene and then seeks a verbal description so that the hearer can guess what object was chosen, or a description game in which the speaker chooses an event or a series of events that took place, conceptualizes a description of these events, and formulates an utterance. The hearer produces a sign of approval when comprehending the utterance and agrees that it describes the scene. An experiment typically focuses on a certain aspect of language, such as evolving a lexicon, or some area of grammar.

Figure 2. Our data in the emergence of language typically involve two robotic 'heads' which observe situations like the one shown here. The robots must use verbal communication to identify objects in the scene or describe events.

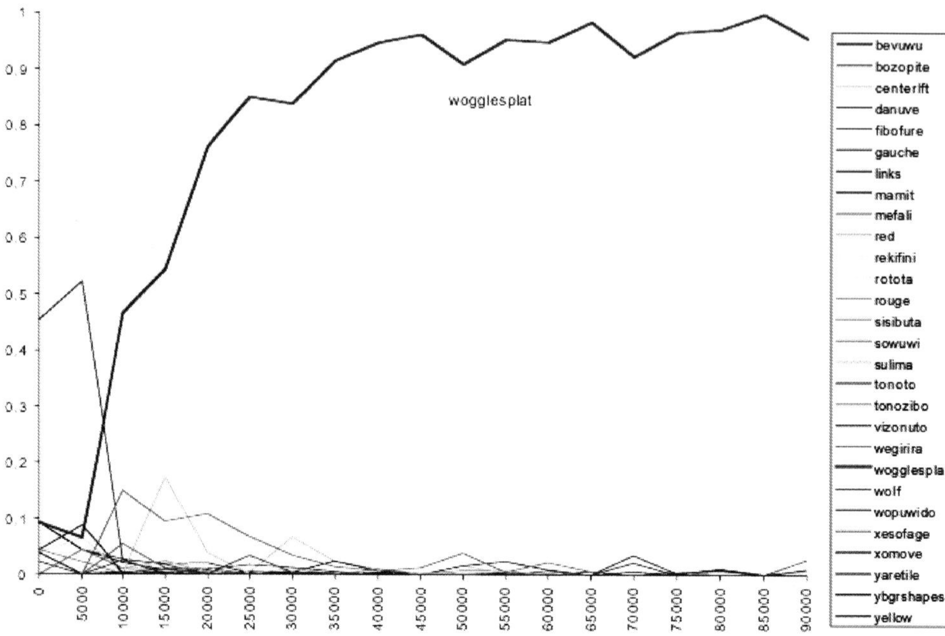

Figure 3.

A meaning-form diagram which graphs for a specific meaning all the possible forms and their frequency of use. A winner-take-all situation is clearly observed. The x-axis shows language games and y-axis the score of forms.

Figure 3 (from Steels *et al.*, 2002) is a typical outcome of an experiment for evolving a lexicon in a steadily growing population reaching 1500 agents towards the end. It shows for a particular meaning, the frequency with which agents use a certain word. The positive feedback between use and success ensures that a winner take all effect takes place: After some time a single word becomes dominant and then propagates in the population, even to new agents that enter the population at regular time intervals.

III: Re-Entrance

1. Rewiring the language system

Now I come to the core issue of this paper. Quite quickly, we were successful in achieving the emergence of a lexicon in our simulations, but the emergence of grammar has turned out to be much more difficult. A breakthrough came as soon as we started to explore the idea that a speaker might apply his own language system (parser and interpreter) to his own utterances, either before transmitting them to the hearer or after observing incomprehension or communicative failure from the part of the listener (figure 4).

It may seem that listening again to a sentence that you produced yourself would not tell you anything new. But that is not the case. The language

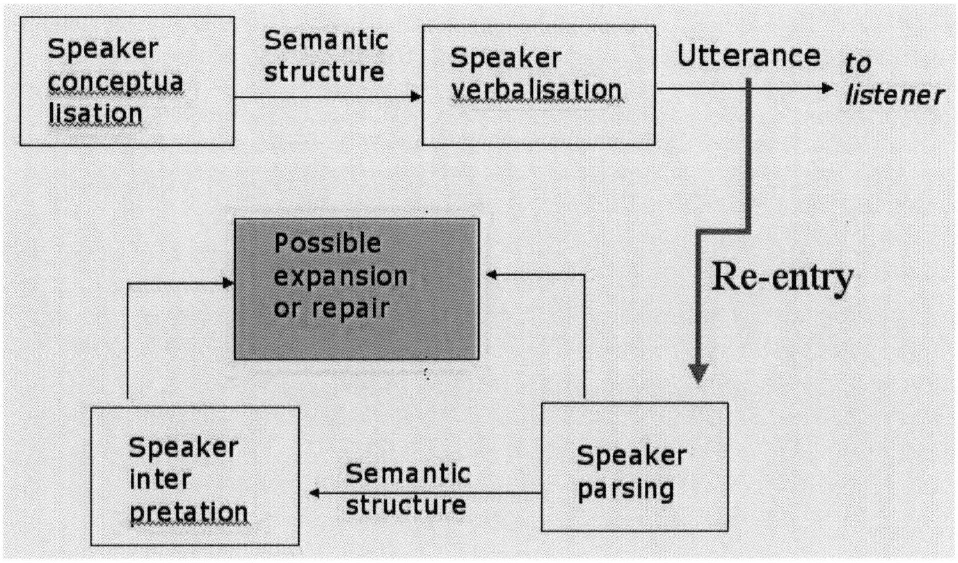

Figure 4.

Output from the language production components is fed back into the language parsing and interpretation components, thus realizing a re-entrant mapping.

production system is very complex and functions automatically and unconsciously. It applies the rules it has and it cannot be planned consciously whether an utterance is going to be adequate from a communicative point of view. The processes of parsing and interpreting the utterance are also automatic and unconscious but they will breakdown or at least detect difficulties if a construction is ambiguous, does not make sense in the present context, or requires more effort than normal. These processes can therefore pinpoint when and where a repair or expansion of the language would be useful and trigger the language invention or language learning subcomponents.

There is plenty of neurophysiological evidence for such a re-entrant system in the human brain. The same left inferior frontal region is activated both for listening to somebody else's speech and for listening to your own speech — even if this speech is not pronounced (McGuire *et al.*, 1996). Moreover this same region has also been observed to be critical for self-awareness (Stuss, 1991), supporting the argument that inner speech contributes to our sense of self (see below). It is also widely known that people self-monitor and correct their own speech on the fly, which suggests that they must simulate to some extent the hearer's interpretation of what they are about to say. Comprehension breakdowns or excessive complexity interrupting the automatic processing of language can even be seen by a shift in brain wave patterns (Brown and Hagoort, 1991).

Auditory hallucinations in schizophrenic patients provide further evidence that a re-entrant language system is possible in the human brain but it also shows that this mechanism may fail to function properly or have negative side effects. Schizophrenic patients claim to hear inner voices. They are no longer able to distinguish whether the speech stream produced by their own language production system is internally or externally produced (Frith and Done, 1988). So as with many evolutionary advantages, there are also disadvantages and possible disorders.

2. Emergent case grammars

Let me now elaborate through a concrete example of grammar, why I believe a re-entrant mapping is necessary for bootstrapping a lexical language into a grammatical one. I will focus on case grammar. It is used in natural languages to make event–object relations explicit. For example, in the sentence *The red block pushed the green block off the table*, English speakers use word order, prepositions and hierarchical structure (Subject–Verb–Object) to indicate that the red block is the active element and the green block undergoes the action. Many languages use other ingredients such as case markers (as in Latin or German) or particles (as in Japanese) to express similar sorts of relations. How might such a case grammar arise and how can it be learned? Suppose that speakers only master a pidgin-like protolanguage, in which there are no case markings nor prepositions, and where word order is not a carrier of meaning. They can then only formulate descriptions like: *Push red block green block.*

When speaker and listener are in the same context and observe the same situation, a unique interpretation might still be found by the listener. However when the listener has to interpret the sentence without being in the same context, or when there are two events in the distant past, one where the red block pushes the green block and another one where the green block pushes the red one, the sentence is ambiguous. The role of case grammar consists in disambiguating such situations. But the question is how can such a case grammar arise?

In one of our experiments (Steels, 2002), the agents start with a pidgin-like language without case marking. When a speaking agent encounters a communicative failure on the part of the listener, he runs the sentence through his own parsing and interpretation processes and discovers where exactly the comprehension failure might have originated. For example, if there are two push events, the speaker's parsing and interpretation processes will detect and signal multiple interpretations. The speaker can then fix this by introducing a marking of the event–object relation, and thus a piece of case grammar into his evolving language.

How can the listener learn the meaning of this marking? First, the listener must identify with enough certainty the interpretation that was intended, for example by additional communications. Next he must fill in the missing gaps by hypothesizing that the unknown marker must be related to his comprehension breakdown. By analysing this breakdown the same way the speaker has done, the meaning of the marker can be deduced. Our simulations (which of course contain much more complexity than discussed here) show that a population of agents programmed with these capabilities indeed evolves a shared system of case markings, signalling in abstract terms the event–object relations occurring in their world.

IV: The Society of Mind

Thus far, I have argued that a re-entrant mapping of speech output as speech input is useful — even necessary — for detecting and repairing language communication, and thus for pushing language and its underlying meaning towards greater complexity. Re-entrant mappings are present all over the brain (Edelman, 1989). For example, the areas in the visual cortex receive much more inputs from higher levels than directly from the retina. It is therefore not surprising that re-entry would also be exploited for language. Language communication is so vital and important that it keeps going even if we do not speak.

Once such a re-entrant system is in place, two other things become possible. First of all, it can be used outside the context of communication. It is possible to conceptualize and verbalize thoughts not to communicate with somebody else but to test out 'how ideas sound', rehearse future dialogue, submit thoughts to self-criticism, or conceptualize and re-affirm a memory of past experiences. This then could have become the foundation for the creation of a sense of self.

Second, language processing components are able to make use of different bodies of grammatical conventions (and this is already achieved in our computer simulations as well). For example, to speak a foreign language a different set of language rules must be activated, even though the overall architecture of language processing remains the same. We store vast amounts of knowledge about how others speak and how they view their world, which makes us capable of conceptualizing a situation and verbalizing it the way somebody else would. As a consequence, we are able to simulate the voice of somebody else, and thus set up internal dialogs which explore different points of view or prepare us for a conversation with somebody else. This is perhaps the way a 'society of mind' as envisioned by Minsky (1985) could have emerged.

At the moment the creative construction and use of representations, both for building a significant model of the self and for preparing interaction with others is completely beyond the current state of A.I. research. In fact, it suggests a whole new avenue of research, which could perhaps build further on the constructionist paradigm which has emerged from Piagetian psychology (Ackermann, 1999) and recent research in cognitive semantics.

V: Conclusions

Normal persons report an 'inner voice' which becomes active when one is not speaking overtly. As argued by many authors, this inner voice is an important part of the experience of the self. I have argued that the technical prerequisites for such an inner voice are quite easily established as soon as there is a re-entrant mapping in which output from speech production is internally streamed as input to speech understanding. I believe that this is a necessary prerequisite for the invention and learning of grammatical language, as our computer simulations have shown. Once this re-entrant mapping is in place, it can also be used as a way to 'listen to oneself', in other words to have an inner voice through which a self-model can be constructed and tested. Although this self-model is not to be identified with consciousness, it is surely part of the conscious experience and therefore the development of complex language communication may have played a crucial role in the origins of consciousness.

Computer simulations and experiments with grounded robots show that it is technically quite feasible to simulate the effect of an inner voice and that it is in fact necessary to achieve certain mental functions such as the ability to acquire complex grammar. However, just as in the case of attentional and turn-taking behaviour, unravelling the information processing foundations of some aspect of consciousness is only part of the full story. The puzzle remains how the information processing structures which realize certain functions associated with consciousness, can give rise to the first-person experience of consciousness (Chalmers, 1996). There is a difference between an actor who displays a certain emotional feeling like anger or love and a person who really experiences these feelings, even though for an outside observer their behaviour might be indistinguishable. Consciousness is not only a matter of information processing (even though that is certainly part of it) but also a felt first-person experience. I personally believe that robotic models may become more and more sophisticated in capturing some of the information processing aspects of consciousness, but whether this will ever lead to the first-person experience itself is a question I would prefer to leave open at the moment.

Acknowledgement

This work was carried out at the Sony Computer Science Laboratoryin Paris. I am indebted to Owen Holland for stimulating the writing of this paper and for the many comments from reviewers. The thoughts expressed here have been partly shaped by discussions with Susan Greenfield, Lewis Wolpert and Daniel Dennett at the Carlo Erba Symposium on Consciousness held in Milan, December 2001. I am grateful to Professor Colonna and Kathleen White for inviting me to this wonderful meeting. Additional ideas have come through my invited talk at the Conference on Consciousness and Language, organized by the Society for the Scientific Study of Consciousness in Barcelona, June 2002, and from discussions with Israel Rosenfield. I am further indebted to Marleen Wynants for comments that improved this paper.

References

Ackermann, E. (1999), 'Enactive representations in learning: Pretence, models and machines', in *Learning Sites: Social and Technological Resources for Learning*, ed. J. Bliss, R. Saljo and P. Light (Amsterdam: Pergamon).

Austin, J.L. (1975), *How To Do Things With Words* (Cambridge, MA: Harvard University Press).

Breazeal, C. (1998), 'A motivational system for regulating human-robot interaction', *Proceedings of AAAI98*, Madison, WI.

Brown, C. and Hagoort, P. (ed. 1991), *Neurocognition of Language* (Oxford: Oxford University Press).

Chalmers, D.J. (1996), *The Conscious Mind: In Search of a Fundamental Theory* (Oxford: Oxford University Press).

Clark, H. and Brennan, S. (1991), 'Grounding in communication', in *Perspectives on Socially Shared Cognition*, ed. L. Resnick, J. Levine and S. Teasley (Washington, DC: APA Books).
Crick, F. and C. Koch (1995), 'Are we aware of neural activity in primary visual cortex?' *Nature*, **375**, pp. 121–3.
Dennett, D. (1991), *Consciousness Explained* (London: Penguin Books).
Edelman, G. (1989), *The Remembered Present: A Biological Theory of Consciousness* (New York: Basic Books).
Fodor, J. (1975), *The Language of Thought* (New York: Crowell).
Frith, C.D. and Done, D.J. (1988), 'Towards a neuropsychology of schizophrenia', *Br J Psychiatry*, **153**, pp. 437–43.
Heine, B. (1997), *Cognitive Foundations of Grammar* (New York: Oxford University Press).
Hopper, (1991), 'Emergent grammar', in Traugott and Heine (1991).
Kuroki, Y., Ishida, T., Yamaguchi, J., Fujita, M. and Doi, T. (2001), 'A small biped entertainment robot', in *Proceedings of the IEEE-RAS International Conference on Humanoid Robots*, Waseda University, Tokyo.
LeDoux, J. (1998), *The Emotional Brain: The Mysterious Underpinning of Emotional Life* (New York: Simon and Schuster).
McGuire, P.K., Silbersweig, D.A., Murray, R.M. David, A.S., Frackowiak, R.S.J. and Frith, C.D. (1996), 'Functional anatomy of inner speech and auditory verbal imagery', *Psychological Medecine*, **26**, pp. 29–38
Minsky, M. (1985), *The Society of Mind* (New York: Simon and Schuster).
Penrose, R. (1990), *The Emperor's New Mind* (Oxford: Oxford University Press).
Rosenfield, I. (1992), *The Strange, Familiar and Forgotten: An Anatomy of Consciousness* (New York: A. Knopf).
Scassellati, B. (1998), 'Imitation and mechanisms of joint attention: A developmental structure for building social skills on a humanoid robot', in *Computation for Metaphors, Analogy and Agents: Vol. 1562 of Springer Lecture Notes in Artificial Intelligence*, ed. C. Nehaniv (Berlin: Springer-Verlag).
Sperber, D. and Wilson, D. (1986), *Relevance: Communication and Cognition* (Cambridge, MA: Harvard University Press).
Steels, L. (2001a), 'Language games for autonomous robots', *IEEE Intelligent Systems*, September/October 2001, pp. 16–22.
Steels, L. (2001b), 'Social learning and verbal communication with humanoid robots', in *Proceedings of the IEEE-RAS International Conference on Humanoid Robots*, Waseda University, Tokyo.
Steels, L. (2002), 'Computer simulations of the origins and evolution of case grammar', Fourth Evolution of Language Conference. Harvard, Cambridge, MA.
Steels, L. and Baillie, J-C. (2002), 'The grounding of dynamic world models by autonomous robots', in *Robotics and Autonomous Systems* (Special Issue on Grounding).
Steels, L. and Kaplan, F. (2001), 'AIBO's first words. The social learning of language and meaning', *Evolution of Communication*, **4** (1), pp.
Steels, L., Kaplan, F.. McIntyre, A. and Van Looveren, J. (2002), 'Crucial factors in the origins of word-meaning', in *The Transition to Language*, ed. A. Wray (Oxford: Oxford University Press).
Stuss, D. (1991), 'Disturbance of self-awareness after frontal system damage', in *Awareness of Deficit After Brain Injury: Clinical and Theoretical Issues*, ed. G.P. Prigatano and D.L. Schacter (New York: Oxford University Press).
Talmy, L. (2000), *Toward a Cognitive Semantics: Concept Structuring Systems (Language, Speech, and Communication)* (Cambridge, MA: The MIT Press).
Tomasello, M. (1999), *The Cultural Origins of Human Cognition* (Cambridge, MA: Harvard University Press).
Traugott, E. and Heine, B. (1991), *Approaches to Grammaticalization. Volume I and II* (Amsterdam: John Benjamins Publishing Company).

William Irwin Thompson

The Borg or Borges?

It is a paradox of the work of Artificial Intelligence that in order to grant consciousness to machines, the engineers first labour to subtract it from humans, as they work to foist upon philosophers a caricature of consciousness in the digital switches of weights and gates in neural nets. As the caricature goes into public circulation with the help of the media, it becomes an acceptable counterfeit currency, and the humanistic philosopher of mind soon finds himself replaced by the robotics scientist. This atmospheric inversion from above to below, one in which a sky turns into the smog of a thickened air, happened once before in the world of knowledge, when Comtian positivism inspired a functionalist approach to the study of the sacred. The social scientists first said that in order to study the sacred, one had to study how it functioned in society; then having contributed to the growth of their own academic domain, they more confidently claimed that what humans worshipped with the sacred was, in fact, their own society. There simply was no such thing as God or the sacred, and so Schools of Divinity began to be eclipsed by the elevation of the new towers of the office buildings of the Social Sciences. Indeed, as I turn now away from my computer screen, I can see outside my window the William James Building of Social Relations competing for dominance of the skyline with the Victorian brick Gothic of Harvard's Memorial Hall.

This clever move to eliminate the phenomenological reality of human consciousness as a prelude to the growth of a new robotics industry is a very successful scam, for it has helped enormously with the task of fund-raising for costly moon shots, such as the Japanese government's 'Fifth Generation Computer Project' which promised to create an autonomously thinking machine in the 1980s. No one seems to talk much anymore about the failure of this project, but the gurus of A.I. continue to prophesy — as Ray Kurzweil now does — that by 2030, humans will be surpassed by machines in cultural evolution.

Both the mechanists and the mystics say that we are now at a great bifurcation in human evolution. The mechanists like Ray Kurzweil, Danny Hillis and Hans Moravec prophesy that we are at the end of the human era, and that 'nanobots' are about to be embedded in our bodies until our antique organs of flesh are entirely surrounded by a new silicon noosphere of networked computers (Kurzweil, 2002;

1999; Moravec, 1988). Like ancient mitochondria or chloroplasts surrounded by the gigantic eukaryotic cell, we are about to be engulfed in the next evolutionary stage. So the mechanists see noetic technologies surrounding human culture and consciousness and compressing it into an endosymbiont in a larger and swifter and more elegant evolutionary vehicle.

Technologists are closer to paranoids than they are to mystics in the sense that they are literalists given to perceptions of misplaced concreteness; they always see spiritual experiences as the products of technology — as emergent domains that are caused by technological innovations, such as LSD or computer networks. The 'difference that makes a difference' — in the famous phrase of Gregory Bateson — between the mystic and the paranoid is that the mystic is in a state of wild cognitive and creative joy, the *satchitananda* of the yogi, but the paranoid is in a condition of anxiety and a cosmic sense of disorientation to the world of spirit that forces him or her into a fixation on literalism and the control of reality through machines. Rather than saying her spiritual intuition has inspired her to see a pattern of connectedness to a world of higher dimensions, she claims to have been abducted by flying saucers who have implanted microchips into her head and are beaming directly into her brain from the mother ship.

Mystics flip this literalism over to see technology as a system of externalized metaphors that derive from pre-existing ontological modes at play and at large in the universe. For them, technology is like the Catholic Baltimore Catechism's definition of a sacrament: 'an outward sign of an inward state'. For the mystic — be she Cabbalist or Sufi — an angel is a 'Celestial Intelligence', a form of cosmic noetic organization that does not require a detour through animal evolution. So when Kurzweil claims that by 2030 implanted nanobots in the bloodstream will enable humans to turn off to the outside world to attune to a virtual reality, the mystic would recognize a literalist rendering of the process of meditation. Kurzweil's vision of the world in 2030 reminds me of Borges' 'Library of Babel': 'I suspect that the human species — the unique species — is about to be extinguished, but the Library will endure: illuminated, solitary, useless, incorruptible, secret' (Borges, 1962). And here we need to be sensitive to the full force of Borges' use of the word 'Babel'.

The mystics, starting with Teilhard de Chardin and Sri Aurobindo in the first half of the twentieth century, also prophesied that we were at a new stage in evolution, but they saw consciousness surrounding technology, and compressing and miniaturizing it into an antique fossil of intermediate cultural evolution as we passed on into a post-human or 'Supramental' era in which we were welcomed back into the cosmic play.

In the eight intensive dimensions that String Theory claims are infolded into the three dimensions of extension and the single dimension of linear time, we now can see that there is more room for humans to think in than we thought we had during the age of 'the conquest of space'. So what Kurzweil conceives of as only possible through the concretization of a machine may actually be possible through a heightened sensitivity to other noetic dimensions. Kurzweil would like the computer to be for him what the organ was to Bach: a way of releasing the human mind into the larger Mind of the universe.

For the mechanists, the flesh is slow, sloppy and wet, and, therefore, primitive. For the Christian mystics, the flesh is the body and blood of the living God. Slow and wet is the ontology of birth and the act of making love. Because the neurons are embedded in an aqueous solution, even distant neurons can participate in a neuronal synchrony through vibrating in the musical harmonies of a single thought. Because the forty Hertz of this neuronal synchrony is slow compared to a silicon computer, it can orchestrate unplanned synchronies in acts of surprise, discovery, analogy, imagination and metaphoric play. Fast is fine for the programmed crystalline world of no surprises and no discoveries, but slow is better for the creative world of erotic and intellectual play.

If one speeds up a Beethoven string quartet, one may enhance the baud rate of data-processing, but one will no longer have music. In fact, with the increase in speed one has lost consciousness of the work. A Beethoven string quartet is, indeed, a rather sophisticated exploration of the nature of time and consciousness, and the interaction between the different instruments is an artistic recapitulation of the evolutionary development of the nervous system in which different channels of information had to be held over in time and cross-referenced with one another to form an 'I'. In Beethoven's 16th Quartet, the third movement, with the markings of '*Lento assai, e cantante tranquillo*' is so slow as to hover at the very edge of melody and silence. Instead of looking to digital computers as a source for metaphors of mind, it would be more instructive to look, or listen, to music. In Kurzweil's emphasis on speed as the unique excellence of mind, he has lived up to his German name too literally, and so, paradoxically, become *langweilig*. The field of consciousness has more to do with slowness and a higher dimensionality, even beyond the three of the physical volume of the brain, in which hyperspheres — or some other higher dimensional topology — involve simultaneity in a neuronal synchrony — in a pattern. A mind, in the opening words of Keats's 'Ode on a Grecian Urn', is a 'still unravished bride of quietness', a 'foster-child of silence and slow time'.

Slowness is fundamental to the nature of consciousness, and here I would define consciousness as the phase-space of the perceptual-motor system. I would argue that in the evolution of consciousness, as far back as the spirochete, it was the delay-space between two different channels of sensory registry, say between light and dark, on the one hand, and acid or base — or a glucose gradient — on the other, that enabled the molecularly lingering traces to be cross-referenced with one another in the formation of an interpretative domain, such as 'Danger!' or 'Flee!'[1] One channel of sensory registration can be a digital gate, a matter of plus or minus, but when two or three differing sensory registrations are cross-referenced to one another, an emergent domain is brought forth. We move up to a new meta-level — like lines forming the higher dimensionality of a cube or hypercube. An interpretive domain is a subjective experience of a sentient being that

[1] I argue this point at greater length in my chapter 'The Past Evolution of Consciousness; from Spirochete to Spinal Chord' (Thompson, 1996/1998, pp. 17–44).

can suffer precisely because it has an identity, and is thus, quite literally, identifying with its sensory registrations in an experiential interpretation of its ontological condition, its life. The neuroscientist Francisco Varela liked to use the Buddhist concept of 'grasping' to mark this aspect of a being identifying so totally with its sensory registrations. As multiple channels of sensory registration develop, a network develops that stabilizes the delay-space, and this is its central nervous system. The natural history of an organism's structural coupling with its environment expresses a reinforcing pattern of response, and this stable response is its identity, its fundamental stabilization of time, its egohood, or, at least, its fundamental *Eigenheit*. If these autonomous identities reproduce themselves with heritable variation over time, we call this evolution.

An engineer can be clever and construct a machine that says 'Ouch!' instead of flashing a red light, but this gnostic demiurge is mimicking consciousness to trick humans. The machine is not a sentient being capable of suffering, and, by imaginative extension and recapitulation of suffering, capable of experiencing compassion for the suffering of other sentient beings.

The mechanists are still not free of the mentality of Galilean Dynamics with its linear system of single causal reductionism. This kind of causal narrative is especially characteristic of the school of Eliminativism of Paul and Patricia Churchland. The simple and linear binary gates of 1 and 0 are fine for artificial neuronal nets and weights, but if one wishes to enfold complexity and make it portable for the life of a unique individual, then the sloppy and chaotic folding of proteins in a cell or of neurons in a brain is the way to go. The brain is actually the most complex small structure we know of in the universe. Like the Borg of *Star Trek,* the mechanists have perverted evolution, for it is the wet and the biological that is the truly advanced design, and our clunky and rigid metalo-plastic computers are the primitive idols of our literal-minded American technoculture. So I side with the mystics and think that the mechanists are caught in the boomerism of American hypercapitalism and are simply hawking their wares.

In this unreflective boomerism of American hypercapitalism, one has to hype one's project to attract venture capital. If one begins to discuss the possible side-effects of the invention, the shadow-side of the design, or the complexity that is the deep background to the object that is being foregrounded, then the investors head for the exits, afraid of law suits. Ironically, this process of self-deception and faulty design only increases the likelihood of lawsuits, for all products have unforeseen side-effects. In our American rush to production and marketing, we take a protein out of context, a gene out of context, a cell out of context, a plant out of an ecology, and a brain out of the context of its complete body incarnation, and we seek ways to sell drugs, genes, patented plants, organs, and soon, perhaps, entire beings. Perhaps Monsanto and Microsoft and Disney will soon be able to effect a merger that will enable them to patent cultures, and EPCOT can take it to the next level. Or could it be that this hostile takeover of culture is what is truly frightening the Muslim world?

The simultaneous fascination and repulsion of Islamic culture to American techno-idolatry is not surprising. Whenever there is a new emergent state of being in the transformations of culture, all of humanity does not immediately shift to the new mentality. If a space voyager wandered around Italy in the fifteenth century, seeking to interview people concerning their excitement at being alive at the time of the Italian Renaissance, most people would not know what the interviewer was talking about. They were still living in the Middle Ages, and would continue to do so until their death.

And so it is now, for most scientists and businessmen are not aware of the implications of complex dynamical systems or of the cultural shift from modernism and the industrial nation-state to planetary culture. So when I am writing about the emergence of a new post-religious spirituality that is in resonance with science — as foreshadowed in such figures as Einstein — I am perfectly aware that I am living in the 'sunset-effect' time of Osama bin Laden and Jerry Fallwell, and that for the billions alive at this moment, their commitment to religion is not about to disappear any time soon. Actually, things need not always disappear in evolutionary extinctions, they can just become surrounded by a new envelopment that is invisible to them. The anaerobic bacteria in my guts are still doing their thing, just as they did billions of years ago before the new atmosphere of oxygen sent them scurrying into the comforts of the dark.

For example, one implication of complex dynamical systems for capitalism is a new version of Adam Smith's 'invisible hand' in which bottom-up causation replaces top-down controls.[2] Both socialism and monolithic corporate capitalism are top-down systems of control that seek to monopolize markets and control governments through lobbying, donations and control of the media. This form of old capitalism is intimately conjoined to modernism, the emergence of the middle class nation-state, and the Galilean Dynamical Mentality. The new capitalism could be a more synergistic system of mutual wealth generation in which groups of inventors bring forth a new cultural ecology. It is a vision of the World Wide Web and the Internet that is more in tune with Linux than Microsoft. Now just as the Inquisition and the Counter Reformation sought to block the Renaissance, so these gigantic corporations like Microsoft or Monsanto are seeking to block the planetary renaissance and this new possibility for capitalism by maintaining the dualistic systems of the domains of the extremely rich and the extremely poor. Microsoft wishes to own the new cultural ecology of the noosphere, and Monsanto, and other companies, are seeking to own the genome of plants, animals and humans. We may slide into a dark age of religious violence with multinational corporations functioning as a tribal amphictyony of competing war lords so that our emergence to an enlightened planetary culture may have to wait a century or two. The Medicis of the Italian Renaissance started out as a merchant class, but they ended up as an aristocracy in the gaudy displays of wealth so characteristic of the baroque economy that was based upon African slavery. The spiritual opening of the Italian Renaissance became blocked by the Inquisition, the Counter Reformation and the Age of Absolutism. Humanity had to wait until the eighteenth century for the Age of Revolution to pick up where the Italian

[2] For an explanation of emergent properties and bottom-up causation in cognitive science, see Thompson & Varela (2001).

Renaissance left off. But if we are lucky, the new form of middle class capitalism that is wed to information technologies and complex dynamical systems may outcompete the recidivist capitalism of the plantations of Monsanto and Microsoft.

At the moment, however, it does not look good, as President Bush *et alia* are wedding imperial capitalism to Christian fundamentalism with its repression of complexity in the arts and sciences in a state of permanent war against terrorists, first foreign, but soon domestic. As Ashcroft has said, carrying on in the tradition of J. Edgar Hoover and Senator McCarthy, those who would restrain him with their misguided liberal notions are 'only giving ammunition to the enemy'.

The boomerism of hypercapitalism that we see expressed in Kurzweil's milenarian vision of the technological replacement of humanity can be easily hitched to Cheney and Rumsfeld's corporate agenda to surround and contain humanity in a perfect system of high- tech defence. Here, it might help to recall that the high philosophic science of Heisenberg and von Weizsäcker joined in with the corporate agenda of I.G. Farben to assist in National Socialism's drive to defeat Bolshevism. And E.O. Wilson's consilient campaign to unify all the sciences also comes at a timely moment to help the Right Wing's desire to eliminate 'secular humanism' (Wilson, 1998). By eliminating philosophical divergence and the distinct cognitive approaches of different disciplines, Wilson's ideological program of elitist unification would serve to remove the humanities and their tradition of liberal humanism in a new scientific version of a Talibanic state of consilient unity. Total explanations soon become totalitarian states. Dissent can be labelled depression and ministered to by the contributions of the pharmaceutical industry. With Ritalin in the schools, Prozac in the universities, Zoloft in the prisons, Ecstasy in the discos, and Viagra in the Senate, America can indeed be at peace with itself to let Kurzweil's machines inherit the Earth.

The cultural evolution of consciousness I had in mind when I coined the phrase 'planetary culture' in more halcyon days was one in which art, science and a post-religious spirituality — like the atmosphere, continents and ocean of a Gaian system — are never unified, but remain free and independent of one another's control, the better to embody complexity and explore the three extensive and the eight intensive dimensions of a universe made out of the music of vibrating strings.

References

Borges, J.L. (1962), 'The library of Babel' in *Labyrinths* (New York: New Directions).

Kurzweil, R. (2002), EDGE [Internet Magazine], March 25.

Kurzweil, R. (1999), *The Age of Spiritual Machines* (New York: Viking).

Moravec, H. (1988), *Mind Children: The Future of Robot and Human Intelligence* (Cambridge, MA: Harvard University Press).

Thompson, E. and Varela, F. (2001), 'Radical embodiment: neural dynamics and consciousness', *Trends in Cognitive Sciences*, **5**, pp. 418–25.

Thompson, W.I. (1996/1998), *Coming into Being: Artifacts and Texts in the Evolution of Consciousness* (New York; St. Martin's Press).

Wilson, E.O. (1998), *Consilience: the Unity of Knowledge* (New York: Vintage Books).